CO2 Capture 2013

Topical Conference at the 2013 AIChE Annual Meeting: Global Challenges for Engineering a Sustainable Future

San Francisco, California, USA
3-8 November 2013

ISBN: 978-1-63439-022-4

Printed from e-media with permission by:

Curran Associates, Inc.
57 Morehouse Lane
Red Hook, NY 12571

CURRAN ASSOCIATES INC. proceedings.com

Some format issues inherent in the e-media version may also appear in this print version.

Printed by Curran Associates, Inc. (2014)

AIChE
3 Park Avenue
New York, NY 10016-5991

Phone: (203) 702-7660
Fax: (203) 775-5177

www.aiche.org

Additional copies of this publication are available from:

Curran Associates, Inc.
57 Morehouse Lane
Red Hook, NY 12571 USA
Phone: 845-758-0400
Fax: 845-758-2634
Email: curran@proceedings.com
Web: www.proceedings.com

TABLE OF CONTENTS

(479a) Functionalization of Metal Organic Frameworks for Enhanced Stability Under Humid Environment for CO2 Capture Applications

Andirova, D.,
Lei, Y., Northeastern University
Choi, S., Northeastern University
Originally presented on: 11/6/2013 12:30:00 - 12:48:00

A need for efficient carbon dioxide capture and storage (CCS) technology heightened in the last few decades due to alarming increase in CO_2 atmospheric concentration which was ascribed to be a possible cause for global climate change. Decarbonization of flue gas from the post-combustion process via adsorption onto porous solid media such as metal organic frameworks (MOFs) have shown their effectiveness among a variety of carbon mitigation technologies.[1,2] Unique structural characteristics of MOFs including extremely large surface area coupled with nanopores, a number of MOFs have demonstrated competitive CO_2 adsorption capacity under the low-pressure conditions simulating the CO_2 concentration in the flue gases (10-15%). However, most of the previous CO_2 capture performance evaluations for MOFs have been carried out using dry CO_2, which is not realistic considering the practical composition of the flue gas. For instance, Mg/DOBDC, the MOF that presents the highest CO_2 adsorption capacity among other MOFs, exhibited significant decrease in its CO_2 capacity under the presence of humidity, possibly due to its strong hydrophilicity.[3]

Here we present the functionalization of MOF frameworks as a facile route to enhance the adsorbent stability and the CO_2 affinity under practical conditions involving humid CO_2 capture. Specifically, the surface of the nanopores in Mg/DOBDC is decorated by the functional groups such as ethylene diamine to prevent the direct contact of water molecules to the open metal sites of the MOF framework, using the post-synthetic functionalization techniques reported in our previous work.[4] The material stability of the functionalized Mg/DOBDC is evaluated via accelerated steam treatment experiments, while being tested by various characterization techniques including surface area and pore size analyzer, XRD, TGA, IR, etc. The CO_2 capture performance of the functionalized Mg/DOBDC upon exposure to humid environment will be also presented.

(1) Choi, S.; Drese, J. H.; Jones, C. W. *ChemSusChem* **2009**, *2*, 796.

(2) Sumida, K.; Rogow, D. L.; Mason, J. A.; McDonald, T. M.; Bloch, E. D.; Herm, Z. R.; Bae, T. H.; Long, J. R. *Chemical Reviews* **2012**, *112*, 724.

(3) Kizzie, A. C.; Wong-Foy, A. G.; Matzger, A. J. *Langmuir : the ACS journal of surfaces and colloids* **2011**, *27*, 6368.

(4) Choi, S.; Watanabe, T.; Bae, T.-H.; Sholl, D. S.; Jones, C. W. *The Journal of Physical Chemistry Letters* **2012**, *3*, 1136.

(479b) Evaluating Different Classes of Porous Materials for Carbon Capture

Obst, J. M., University of California-Berkeley
Lin, L. C., University of California-Berkeley
Berger, A. H., Electric Power Research Institute
Martin, R. L., Lawrence Berkeley National Laboratory
Haranczyk, M., Lawrence Berkeley National Laboratory
Bhown, A. S., Electric Power Research Institute
Reuter, K., Fritz-Haber-Institut der Max-Planck-Gesellschaft
Smit, B., University of Amsterdam
Originally presented on: 11/6/2013 12:48:00 - 13:06:00

Carbon Capture and Storage (CCS) is one of the promising ways to significantly reduce the emission of CO_2 from power plants. The separation and compression of CO_2, however, is currently considered the most energy consumptive part. It is therefore important to find adequate materials to reduce the energy load additionally imposed on a power plant, which necessitates the development of a suitable metric to rank those materials. The concept of parasitic energy is a potential means for the evaluation of CCS materials. By combining thermodynamic properties and optimizing the process conditions the parasitic energy requirement can be determined on a material-by-material basis. Ranking materials according to their energy efficiency in CCS not only provides a direct comparison to current state-of-art technologies like MEA but also enables the application to different capturing strategies.

Over 50 different materials, including metal-organic frameworks (MOFs), zeolitic imidazolate frameworks (ZIFs), porous polymer networks (PPNs), zeolites, and hypothetical materials were analyzed using the approach of parasitic energy and compared to the results of other suggested ranking strategies such as selectivity and working capacity. Furthermore, the influence of different regeneration techniques as well as the impact of the adopted prediction model for mixture adsorptions on the parasitic load were investigated.

(479c) Carbon Dioxide Separation Using Highly Selective Narrow Pore Zeolite ZK-4

Cheung, O., Stockholm University
Bacsik, Z., Stockholm Univerisity
Mace, A., Stockholm Univerisity
Krokidas, P., Stockholm University
Laaksonen, A., Stockholm Univerisity
Hedin, N., Stockholm Univerisity
Originally presented on: 11/6/2013 13:06:00 - 13:24:00

In recent years, zeolites were comprehensively studied for their potential as CO_2 sorbents in carbon capture and storage (CCS) processes. Zeolites offer high CO_2 capacity, selective CO_2 sorption, high thermal stability and low costs. Zeolite ZK-4, is a silicon rich version of a zeolite, which has the same framework topology as the Zeolite A. Previously, we studied and showed that the Zeolite NaKA could be tuned into a highly CO_2 selective sorbent. ZK-4 with its potential for high Si/Al ratios could potentially alleviate the disadvantages linked to the very hydrophilic nature of zeolite A. Here, we show that a zeolite ZK-4 (Si/Al =1.36) could be tuned into a highly CO_2 selective form by carefully tuning the cation ratio (K^+/(K^++Na^+))

When the cation ratio of K^+ content was over ~ 25 atomic %, the sorbent exhibited a molecular sieving of CO_2 over N_2 and was highly selective towards CO_2 (ideal CO_2/N_2 selectivity >200). We ascribe this manifold enhanced selectivity to the specific location of the K^+ cation in the pore windows, where their comparably larger size hinders the diffusion of N_2 more effectively than for CO_2. The CO_2 sorption capacity remained high on the CO_2 selective materials (~ 4.1 mmol/g 273K, 101 kPa). The heat of CO_2 physisorption varied between 32 – 45 kJ/mol and was partly related to the K^+ ion content. The rate of CO_2 physisorption was relatively fast. Infrared spectroscopy and molecular modeling gave an insight into the configuration of adsorbed CO_2 and how the cations affected the sorption of CO_2. In summary, zeolite ZK-4 offers the similar CO_2 adsorption characteristics as zeolite NaKA, but its more hydrophobic nature and the possibility for varying the Si/Al ratio allow higher flexibility for tuning this material for specific application. We are continuing these studies by increasing the Si:Al ratio of ZK-4 until the molecular sieving capacity of CO_2-over-N_2 is lost.

(479d) Understanding Carbon Dioxide Adsorption On Alkali Metal Exchanged Zeolite SSZ-13
Pham, T., University of Delaware
Lobo, R. F., University of Delaware
Originally presented on: 11/6/2013 13:24:00 - 13:42:00

Understanding Carbon Dioxide Adsorption on Alkali Metal Exchanged Zeolite SSZ-13

Trong D.Pham[1], Matthew R. Hudson[2], Craig M. Brown[2], Raul F.Lobo[1*]

[1] *Center for Catalytic Science and Technology, University of Delaware, Department of Chemical and Biomolecular Engineering, USA, lobo@udel.edu.*

[2] *Center for Neutron Research, National Institute of Standards and Technology, Gaithersburg, Maryland 20899-6102, USA.*

Keywords: CO_2, SSZ-13, adsorption, neutron diffraction

Introduction

The emission of anthropogenic carbon dioxide into the atmosphere due to the combustion of fossil fuels is one of the main causes of climate change. Low silica zeolites (types A, 13X, zeolite Y) are among the most commonly used CO_2 adsorbents in industrial gas separations[1,2], but are generally strong adsorbents, making regeneration difficult [3]. Because of lower Al content, high silica zeolites with a less hydrophilic pore environment are an interesting option with respect to the regeneration and the tolerance to polar impurities in the gas such as water vapor and sulfur compounds[4]. Here we report an investigation of CO_2 adsorption on high-silica zeolites SSZ-13 (CHA framework-type, Si/Al=6 and 12), examining the effects of framework Si/Al ratio and extra-framework cations (Li^+, Na^+, K^+) on CO_2 adsorption isotherms and heats of adsorption. The interaction of CO_2 with the SSZ-13 is investigated at the molecular level using Rietveld refinement of neutron diffraction patterns of adsorbed CO_2 in the host zeolite.

Experimental

SSZ-13 samples (Si/Al=6 and 12) were prepared by hydrothermal methods using N,N,N-Trimethyl-1-adamantanammonium ion as a structure-directing agent. Element compositions of synthesized and ion-exchanged samples were determined by Energy-dispersive X-ray spectroscopy (EDX). CO_2 adsorption isotherms of the SSZ-13 samples at various temperatures (273, 303 and 343 K) were measured on a Micromeritics ASAP 2020 device up to ambient pressure. Neutron powder diffraction (NPD) measurements were conducted on the high-resolution diffractometer BT1 at NIST, using a Ge(311) monochromator, λ = 2.0787(2) Ã… for bare and CO2 gas-dosed alkali-exchanged SSZ-13.

Results and Discussion

Na-SSZ-13/6 is remarkable because it shows comparable equilibrium CO_2 capacity to 13X and Y, and higher capacity than NaA, three zeolites with much higher aluminum content (Table 1).Â This is because *all* the cations in the smaller cages of SSZ-13 are accessible to CO_2, and despite the fact that SSZ-13 has a lower cation concentration compared to other zeolite frameworks. Na-SSZ-13/12 adsorbed less CO_2 than the lower Si/Al samples under the same conditions, as expected, but shows much higher adsorption capacity than FER and STI with similar Si/Al ratio.Â This difference is attributed to the higher microporous volume of the SSZ-13 sample.

Table 1. Adsorption capacity of CO_2 on various sodium-exchanged zeolites at 303 K and 1atm.

Samples	Na-13X	NaY	NaA	Na-FER[6]	Na-STI [6]	Na-SSZ-13/6	Na-SSZ-13/12
Si/Al ratio	1.25	2.47	1	9	15	6	12
Uptake (mmol/g)	5.23	5.03	4.13	2.6	2.5	5	3.96

*ref [6]: The experiment was carried out at 298K and 750torr

Figure 1: Schematic view of cation positions in SSZ-13 framework (left) and CO_2 positions in Na-SSZ-13 (Si/Al~12) (right)

Rietveld refinement of neutron diffraction on bare adsorbent SSZ-13/12 shows that Li^+, Na^+ is located right above the center of the hexagonal prism (site SII), and the larger cation K^+ is in the middle of 8-membered rings. With 4 K^+ cations/cage in K-SSZ-13/6, K^+ could reduce the effective adsorption capacity of the zeolite by blocking access to some of the SSZ-13 cavities. Difference Fourier maps were used to locate the positions of CO_2 molecules in the zeolite. At a loading of 0.66 molecules CO_2 per 8-ring window on Na-SSZ-13/12, we found that CO_2 is located in the center of the 8-membered ring window instead of directly interacting with extra-framework cations, which is similar to the results reported by Hudson et al[5] for Cu-SSZ-13. Na^+ is bound to 3 oxygen atoms of the 6-ring with a bond length of 2.35 Ã…, and the closest distance of O(CO_2)-Na is 3.71 Ã…. This distance is too long for efficient interaction coupling of the CO_2 and the cations suggesting that adsorption is driven by the electrostatic interactions of the framework oxygen atoms in the 8-ring and the carbon atom in CO_2.

Conclusions

SSZ-13 has a high potential for CO_2 capture, reaching an adsorption capacity of 5.1 mmol CO_2 per gram of zeolite at ambient temperature and pressure, a value comparable with the best-known zeolite adsorbents (X, Y). The adsorption characteristics of SSZ-13 are strongly dependent upon the cation concentration, the cation-type, and the CO_2 loading amount.

Acknowledgements

This work was supported by the National Science Foundation through grant CBET-940768 and National Institute of Standards and Technology Center for Neutron Research (NCNR). Trong Pham acknowledges support from Vietnam Education Foundation (VEF). MRH acknowledges support from the NRC Postdoctoral Research Fellowship.

References

[1] Yang, H.; Xu, Z.; Fan, M.; Gupta, R.; Slimane, R. B.; Bland, A. E.; Wright, I., *Journal of Environmental Sciences-China* **2008,** *20* (1), 14-27.

[2] Walton, K. S.; Abney, M. B.; LeVan, M. D., *Microporous Mesoporous Mater.* **2006,** *91* (1-3), 78-84.

[3] Harlick, P. J. E.; Tezel, F. H., *Microporous Mesoporous Mater.* **2004,** *76* (1-3), 71-79.

[4] Bao, Z.; Yu, L.; Dou, T.; Gong, Y.; Zhang, Q.; Ren, Q.; Lu, X.; Deng, S., *J. Chem. Eng. Data,* **2011,** *56* (11), 4017-4023.

[5] Hudson, M. R.; Queen, W. L.; Mason, J. A.; Fickel, D. W.; Lobo, R. F.; Brown, C. M., *J. Am. Chem. Soc.* **2012,** *134* (4), 1970-1973.

[6] Zukal, A.; Zones, S. I.; Kubu, M.; Davis, T. M.; Cejka, J., *Chempluschem* **2012,** *77* (8), 675-681. ADDIN EN.REFLIST

(479e) Molecular-Level Insight Into Unusual Low Pressure CO2 Affinity in Pillared Metal-Organic Frameworks

Burtch, N. C., Georgia Tech
Jasuja, H., Georgia Institute of Technology
Dubbeldam, D., University of Amsterdam
Walton, K. S., Georgia Institute of Technology
Originally presented on: 11/6/2013 13:42:00 - 14:00:00

Fundamental insight into how low pressure adsorption properties are affected by chemical functionalization is critical to the development of next generation porous materials for post-combustion CO_2 capture. In this work, we present a systematic approach to understanding low pressure CO_2 affinity in isostructural metal-organic frameworks (MOFs) using molecular simulations and apply it to obtain quantitative, molecular-level insight into interesting experimental low pressure adsorption trends in a series of pillared MOFs. Our experimental results show that increasing the number of non-polar functional groups on the benzene dicarboxylate (BDC) linker in the pillared DMOF-1 $[Zn_2(BDC)_2(DABCO)]$ structure is an effective way to tune the CO_2 Henry's coefficient in this isostructural series. These findings are contrary to the common scenario where polar functional groups induce the greatest increase in low pressure affinity through polarization of the CO_2 molecule. Instead, MOFs in this isostructural series containing nitro, hydroxyl, fluorine, chlorine, and bromine functional groups result in little increase to the low pressure CO_2 affinity. Strong agreement between simulated and experimental Henry's coefficient values is obtained from simulations on representative structures, and a powerful yet simple approach involving the analysis of the simulated heats of adsorption, adsorbate density distributions, and minimum energy 0 K binding sites is presented to elucidate the intermolecular interactions governing these interesting trends. Through a combined experimental and simulation approach we demonstrate how subtle, structure-specific differences in CO_2 affinity induced by functionalization can be understood at the molecular-level through classical simulations. This work also illustrates how structure-property relationships resulting from chemical functionalization can be very specific to the topology and electrostatic environment in the structure of interest. Given the excellent agreement between experiments and simulation, predicted CO_2 selectivities over N_2, CH_4, and CO are also investigated to demonstrate that methyl groups also provide the greatest increase in CO_2 selectivity relative to the other functional groups. These results indicate that methyl ligand functionalization may be a promising approach for creating both water stable and CO_2 selective variations of other MOFs for various industrial applications.

(479f) Adsorption of CO2 Over Supported Amine Adsorbents in the Presence of SOx and NOx Impurities

Rezaei, F., Georgia Institude of Technology
Jones, C. W., Georgia Institute of Technology
Originally presented on: 11/6/2013 14:00:00 - 14:18:00

For supported amine adsorbents to be useful in practical CO_2 capture application, it is crucial to exhibit a stable performance and a long lifetime in the presence of flue gas impurities such as SOx and NOx. Here, the stability of supported aminosilica adsorbents to SO_2 and NOx impurities in CO_2 capture process is evaluated by conducting equilibrium and dynamic adsorption experiments. The effect of temperature, feed concentration and adsorbent amine loading are investigated by performing the adsorption experiments in TGA and fixed bed using different SO_2, NO and NO_2 concentrations. In equilibrium measurements, the single-component adsorption of CO_2, SO_2, NO and NO_2 onto supported amine adsorbents is carried out by directly exposing the material to the various gases in carrier gas. The CO_2 capacity is measured before and after exposure to SO_2, NO or NO_2 to evaluate any CO_2 capacity loss due to irreversible binding of SO_2, NO or NO_2 to the supported amine adsorbents. Furthermore, to assess any synergies in co-adsorption of SOx or NOx and CO_2 gases, three gas mixtures comprising SO_2/CO_2, NO/CO_2 and NO_2/CO_2 are used in addition to single-component adsorption measurements. The experiments are performed by simultaneously exposing the adsorbents to CO_2 and SO_2 or CO_2 and NO/NO_2 and thereby their CO_2 uptake is measured at different temperatures.

(479g) Selective Dynamic CO2 Separations On Mg-MOF-74 Pellets At Low Pressures: A Detailed Comparison With 13X

Remy, T., Vrije Universiteit Brussel
Peter, S. A., Indian Institute of Technology Bombay
Van Der Perre, S., Vrije Universiteit Brussel
Baron, G., Vrije Universiteit Brussel
Denayer, J., Vrije Universiteit Brussel
Originally presented on: 11/6/2013 14:18:00 - 14:36:00

Selective separation of CO_2 from various gas streams is typically carried out via chemical absorption with amine solutions in the (petro-) chemical industry. Adsorption with porous solids such as metal-organic frameworks (MOFs) is an interesting alternative because this technique could be more energy efficient and thus cost reducing.

Different MOFs from the MOF-74 or MDOBDC family have large CO_2 capacities when exposed to pure CO_2 streams (M = Ni or Mg / DOBDC = 2,5 dioxido-1,4-benzenedicarboxylate). These MOFs contain open metal sites and experiments with small amounts of material indicate a large selectivity and capacity for CO_2 [1-2].

In this work, the potential of the promising Mg-MOF-74 material was studied in closer detail and compared to the benchmark zeolite 13X. Adsorption isotherms of CO_2, CH_4 and N_2 were recorded on MOF powder using the gravimetric technique at different temperatures (273 – 373 K). The uptake measurements indicate much lower affinity for CH_4 and N_2 and suggest a larger capacity and selectivity for the MOF in relevant low-pressure CO_2 separations.

To assess the dynamic separation potential we developed a new pellet preparation technique to ensure a minor capacity loss upon shaping of the powder. Breakthrough experiments with mixtures of CO_2 and CH_4 on a column packed with Mg-MOF-74 pellets at 308 K and 1 bar indicated a higher CO_2 capacity and separation efficiency for the MOF with respect to the benchmark 13X zeolite. The MOF regains 81% of its original capacity when purging with helium at 308 K during 10 minutes and can be fully regenerated at 353 K, which indicates the facile desorption of CO_2 on this material. However, the performance of Mg-MOF-74 severely deteriorates upon long-term exposure to relevant impurities in CO_2 separations at low pressures such as water and oxygen.

[1] P.D.C. Dietzel, V. Besikiotis, R. Blom, J. Mater. Chem., **19**, (2009).

[2] R. Krishna, J.M. van Baten, J. Membr. Sci., **377**, (2011).

(479h) Metal-Organic Framework With Optimal Adsorption Thermodynamics and Kinetics for CO2 Separations

Belmabkhout, Y., King Abdullah University of Science and Technology

Originally presented on: 11/6/2013 14:36:00 - 14:54:00

Metal-Organic Framework with Optimal Adsorption Thermodynamics and Kinetics for CO_2 Separations

Youssef Belmabkhout[1], Amy Cairns[1], Ryan Luebke[1], Patrick Nugent[2], Stephen D. Burd[2], Michael. J Zaworotko[2] and Mohamed Eddaoudi[1,2].

[1]KAUST Advanced Membranes & Porous Materials Center, Functional Material Design, Discovery and Development (FMD[3])King Abdullah University of Science and Technology
4700 King Abdullah University of Science and Technology Thuwal 23955-6900 (Kingdom of Saudi Arabia)

[2]Department of Chemistry, University of South Florida, 4202 East Fowler Avenue, Tampa, Florida 33620, USA.

**Corresponding author email: youssef.belmabkhout@kaust.edu.sa*

Metal Organic Frameworks MOFs (MOFs) is burgeoning class of materials are comprised of metals or metal clusters ("nodes") coordinated to multi-functional organic ligands ("linkers") and they offer myriad of surface areas up to 7000 m^2/g [1]. Furthermore, the modular nature of MOMs and their use of known coordination chemistry lead to enormous diversity of structures and properties [2]. Unfortunately, though extra-large surface area and pores facilitate high gravimetric uptake of gases at low temperature and/or high pressure, they are not conducive to highly efficient separations at the relatively low pressures necessary for the capture of CO_2 particularly for post-combustion. Unsaturated metal centers (UMCs) [3] or organic amines [2] that chemically interact with CO_2 can promote enhanced binding for CO_2 but there are drawbacks: high energy costs associated with activation, regeneration and recycling of the sorbent material, especially for amines [2]; competition with water vapor, especially for UMCs [4]; selectivity tends to monotonically decrease with increased loading of sorbate. Consequently, a sorbent with favorable CO_2 sorption kinetics and/or constant and moderate adsorption enthalpies over a wide range of CO_2 loading will permit efficient CO_2 capture from flue gas with low regeneration costs. We describe herein an approach to synergistically balance the kinetics and thermodynamics of CO_2 sorption selectivity through a crystal engineering and reticular chemistry strategy that affords control over pore functionality and size. The resulting MOFs are microporous sorbents with narrow one dimensional square channels lined by periodically arrayed SiF_6^{2-} (SIFSIX) anions. These MOFs exhibit exceptional CO_2 selectivity under conditions practically relevant to CO_2 separation and capture in the context of post-combustion (flue gas CO_2/N_2), pre-combustion (shifted synthesis gas stream CO_2/H_2) and natural gas upgrading (natural gas cleanup CO_2/CH_4) [5].

[1] Farha, O. K. *et al. J. Am. Chem. Soc.* 134, 15016-1502 (2012).

[2] Sumida, K. *et al. Chem. Rev.* 112, 724-781 (2012).

[3] Caskey, S. R., Wong Foy, A. G. & Matzger, A. J.. *J. Am. Chem. Soc.* 130, 10870-10871 (2008).

[4] Kizzie, A. C., Wong Foy, A. G. & Matzger, A. J. *Langmuir* 27, 6368-6373 (2011).

[5] Nugent, P., Belmabkhout, Y., Burd, S. et al. Nature (2013) 10.1038/nature11893.

(524a) Dynamic Modelling and Analysis of a Coal Fired Power Plant Integrated With Post-Combustion CO2 Capture Process

Mac Dowell, N., Imperial College London
Shah, N., Imperial College London
Originally presented on: 11/6/2013 15:15:00 - 15:40:00

There have been several design and optimisation studies in the context of amine-based post-combustion CO_2 capture (PC-CC) from coal-fired power stations, and it is well accepted that the principal contributor to the capital-cost (CAPEX) of amine-based PC-CC is the absorber, with the majority of operating cost (OPEX) being accounted for by solvent regeneration[1]. However, the majority of these studies consider only the maximal stable generation (MSG) point of the power station. In reality, many power-plants do not operate consistently at the MSG. Thus, the cost optimal design of a PC-CC process must account for the dynamic operation of many fossil-fuel power-plants. The economic operation of a decarbonised fossil-fired power-plant is a complex trade-off between the time varying cost-of-electricity and cost of fuel. These variables will in turn affect the cost-optimal degree of CO_2 capture, and thus the design and operation of the CO_2 capture plant.

In this work, we present a new approach for the multi-period design of an integrated coal-fired power-plant and an alkanolamine-based CO_2 capture process. A validated[2] rate-based model of an absorber/stripper system for the chemisorption of acid gas in aqueous solvent solutions is used to represent the CO_2 capture process. To account for the non-idealities that are typical of amines and water, the statistical associating fluid theory for potentials of variable range (SAFT-VR)[3],[4] is used. This is a molecular approach, specifically suited to associating fluids. The SAFT formalism is used to represent some of the equilibrium reactions characterising the system, thereby simplifying the description of the chemical reactions[5]. The dynamic operation of the power-plant is described using data obtained from a real power station describing the variation in fuel burn rates in response to changing power demand. All models are implemented in the gPROMS[6] software package.

Cost-optimal process design and operation are identified via a constrained, multi-parametric optimisation approach. The objective function is the minimisation of total process cost (both CAPEX and OPEX) per unit of CO_2 captured. The design variables are the flowrate and state of the lean solvent stream as well as column geometry. Key performance indicators such as CO_2 emissions, solvent losses to the environment as well as the minimisation of energy required for solvent regeneration and also minimisation of CO_2 emissions to the atmosphere are included in the optimality criteria.

[1] Mac Dowell, N. et al, Energy. Environ. Sci., 3, 2010, 1645-1669

[2] Mac Dowell, N., Samsatli, N. & Shah, N. Int. J. GHG. Con., 2013, 12, 247-258

[3] Chapman, W.G., Gubbins, K.E., Jackson, G. & Radosz, M., Ind. Eng. Chem. Res. 29, 1709-1721., 1990

[4] Gil-Villegas, A., Galindo, A., Whitehead, P. J., Mills, S. J. & Jackson, G., J. Chem. Phys. 106 (10), 1997

[5] Mac Dowell, N., Llovell, F., Adjiman, C. S., Jackson, G and Galindo, A., Ind. Eng. Chem. Res., 49(4), 1883-1899, 2010

[6] Process Systems Enterprise (PSE) Ltd. http://www.psenterprise.com/index.html

(524b) Dynamic Modeling Studies On Absorption of CO2 in Rotated Packing Bed

Kang, J. L., National Tsing Hua University
Kai, S., Qilu university of technology
Wong, D. S. H., National Tsing Hua University
Jang, S. S., National Tsing Hua University
Tan, C. S., National Tsing Hua University
Originally presented on: 11/6/2013 15:40:00 - 16:05:00

Absorption by chemical solvents is one of the most mature technologies developed for CO2 capture from flue gas. Traditional devices, such as packed towers, required huge volume because of mass transfer limitation□which is a critical factor contributing to the cost of the process. The rotating packed bed (RPB) was proposed to relax mass transfer limitations Lin et al 2003, Chen and Tan 2006, Jassim et al. 2007, Cheng and Tan 2009, 2011. However, there are a limited number of studies on modeling of CO2 capture by alkanoamines using RPB. For example Yu et al 2012 proposed a model with an assumption of 6 stirred tanks in series followed by a gas–liquid contactor.

In this work, a model of absorption of CO2 by mono-ethanolamine (MEA) in RPB was developed using two-film theory for mass transfer. Traditionally, in RPB studies, the Onda's correlation (Onda et al 1968) was used for predicting gas phase mass transfer coefficients. The correlation proposed by Tung and Mah 1985 was used for predicting liquid phase mass transfer coefficients. The gas-liquid interfacial area was predicted by the Onda's correlation. The only dependence on gravity is in the gas-liquid interface area correlation, which depends on the Froude number (Fr). However the Onda correlation has only an almost negligible dependence upon Fr ($Fr^{-0.05}$); leading to under prediction of capture efficiency. Recently, Hanley and Chen (2012), in a new study on mass transfer characteristics of trays and packing, proposed a new correlation for the gas-liquid interfacial area, which has a more prominent dependence on gravity ($Fr^{-0.161}$). We found that this modified expression can lead to better agreement with experimental data using 30 wt% MEA solutions without the need to introduce an extra void volume.

In CO2 capture by alkano-amine solutions, mass transfer is accompanied by reaction. A pseudo-first order reaction constant was used to calculate enhancement factor of mass transfer rate. In this study, we found that this model significantly underestimated capture efficiency for experiments conducted with high concentrations of MEA. Aboudheir et al 2003 suggested that pseudo-first order reaction is not appropriate and suggested a new kinetic model to account for improved enhancement at high MEA concentrations. However, the model showed some systematic errors with respect to MEA concentrations. An empirical correction was introduced in this work to improve the prediction of enhancement factors. These corrections of enhancement factors enable us to predict experiment data provided by Jassim et al 2007 at high MEA concentrations.

In summary, an improved model of CO2 capture by MEA solutions by RPB was developed by modifying the gravity dependence of gas-liquid interfacial area and the enhancement factor at high MEA concentrations. By integrating these improvements obtained from independent studies, we can substantially improve the predictions of CO2 capture efficiency provided in the literature by various research groups.

(524c) A Plantwide Dynamic Model of Post-Combustion Amine Scrubbing With Aqueous Piperazine

Walters, M. S., The University of Texas at Austin
Edgar, T. F., The University of Texas at Austin
Rochelle, G. T., The University of Texas at Austin
Originally presented on: 11/6/2013 16:05:00 - 16:30:00

Dynamic process models are important for designing control strategies and optimizing off-design operating conditions. An equation-based dynamic model has been developed for a post-combustion amine scrubbing process including CO_2 absorption, CO_2 desorption, and heat exchange. Aqueous piperazine was used as the amine solvent for CO_2 absorption in this model. Piperazine is a second generation solvent that has superior energy performance compared to the first generation solvent monoethanolamine, which was used in previously reported first principle dynamic modeling works. Unit level dynamic models for alternative process configurations, which have been shown to reduce capital and operating costs compared to the simple absorber/stripper base case, were developed here. For the absorber, an intercooling option was included to increase mass transfer by decreasing the temperature bulge that occurs as a result of the exothermic absorption of CO_2. An advanced flash stripper configuration with CO_2-rich solvent bypass options for tighter energy integration was used in place of a simple stripper. The increased energy recycle from advanced configurations creates more complicated dynamics and added complexity. Steady state representations were used for other unit operations, including heat exchangers, pumps, and control valves.

The individual unit models were combined into one flowsheet. This plantwide dynamic model was implemented in gPROMS©, an advanced process modeling platform. The model was validated using pilot plant data from the Separations Research Program at the University of Texas at Austin. In order to validate the model with pilot plant data, reasonable systematic adjustments to CO_2-loading measurements and heat loss estimates were required. After the adjustments were made, validation with temperature, pressure, and CO_2-loading measurements observed at the pilot plant had adequate agreement to the predicted model outputs. Both steady state runs and dynamic behavior between runs were used for validation purposes. The multiple time scale behavior of this process was demonstrated by showing that the ratio of the amount of solvent recycled to the amount of CO_2 and solvent makeup continuously added to the system greatly exceeds unity. A process control hierarchy is proposed based on the observed fast time scale associated with the large recycle flows and the slow time scale associated with total energy and CO_2 holdups in the system.

(524d) Study of the Effect of Stripping Components in Absorption Processes for CO2-Removal From Power Plant Flue Gases

Kossmann, A., Institute of Plant and Process Technology
Moser, P., RWE AG
Klein, H., Institute of Plant and Process Technology
Originally presented on: 11/6/2013 16:30:00 - 16:55:00

One possibility for removal of CO_2 in conventional coal-fired power plants is the so-called Post-Combustion-Capture process. In this process CO_2 is separated from the power plant flue gas by absorption. Amongst others aqueous amine solutions are considered as one possible solvent for absorption. Advantageous of the Post-Combustion-Capture process is the fact that existing power plants could be retrofitted. A disadvantage however is the additional energy consumption occurring when regenerating the solvent in the desorber. For regeneration steam is taken from the turbines of the power plant. This reduces the overall efficiency of the power plant significantly.

Adding an organic component, which is almost immiscible with water, could have a positive effect on the regeneration of the solvent and therefore also on the overall efficiency of the power plant. By adding this organic component a miscibility gap is formed in the desorber and subsequently the boiling temperature decreases, an effect also known from steam distillation. As the temperature in the desorber is lower, the pressure of the required steam for regenerating the solvent can also be lower.

In this work the influence of the miscibility gap on the absorption process is examined. Therefore, a model for a three phase equilibrium was developed. Based on this model for phase equilibrium a multistage column model, representing the desorber, was designed. With this model and another model for the absorber, both implemented in MATLAB, the overall absorption process can be simulated.

First results show that depending on the choice of the organic component the boiling temperature of the reboiler can be significantly lowered. The effect on the overall performance of power plant including CO_2 separation was then estimated by a simple exergetic analysis. The considered process options showed about the same performance in this estimation. Therefore, a more detailed evaluation including a power plant simulation model is planned for the near future. Furthermore, other amines and amine strength will be subject of further optimizations.

(524e) Incorporating Technology Improvements in Amine Scrubbing in Power Plants With Post-Combustion Capture

Lucquiaud, M., The University of Edinburgh
Liang, X.,
Gibbins, J., University of Edinburgh
Originally presented on: 11/6/2013 16:55:00 - 17:20:00

Risks associated with technology, market and regulatory uncertainties for First-Of-A-Kind fossil power generation with CCS can be mitigated through innovative engineering approaches that will allow solvent and process developments in amine scrubbing technologies be subsequently incorporated into the next generation of CCS plants.Power plants capable of improving their economic performance will benefit financially from being able to upgrade their solvent technology or process.

This work identifies the key requirements to upgrade solvents used in amine scrubbing technologies by examining both the scrubbing process and its integration with the power cycel. A sensitivity of solvent properties is undertaken to identify bottleneck and quantify lock-in related to the overall performance of the power plant.

One of the most important requirements for upgradeability is for the base power plant to be able to operate with any level of steam extraction and also with any level of electricity output up to the maximum rating without capture. This requirement will also confer operational flexibility and so is likely to be implemented in practice on new plants or on any integrated CCS retrofit project.

(524f) CO2 Post Combustion Capture Operational Flexibility and Scale-Up For Power Plants
Singh, P., IEAGHG
Originally presented on: 11/6/2013 17:20:00 - 17:45:00

CO_2 Post Combustion Capture Operational Flexibility and Scale-up for Power Plants

Prachi Singh, John Davison

International Energy Agency Green House Gas R&D Programme (IEAGHG), The Orchard Business Centre, Stoke Orchard, Cheltenham, GLOS UK, GL52 7RZ, UK

<>1.Introduction

Several governments and international organizations have set 2020 as the target for broad commercial deployment of Carbon Capture and Storage (CCS). With this target in mind, it is clear that the initial commercial and full scale capture plants will have to be based on currently available technologies. These commitments and agreed targets create an important role for the solvent-based post-combustion capture technology, which is considered to be the most mature of the capture technologies available today. CO_2 Post combustion capture technology provides a retrofit possibility and is already available on relatively small industrial scale. Moreover now a days existing and new power plants must face the challenges of the liberalized electricity market and the requirement to cover intermediate and peak load constraints, in order to respond to the daily and seasonal variation of the electricity demand. Most studies undertook CO_2 post combustion capture at power plant base load operation, but it is now clear that CO_2 post combustion capture technology will also need to be able to respond to the requirements of the new liberalized electricity market. Therefore, in this work, current capabilities of conventional Ultra Supercritical Pulverised Coal (USCPC) and Natural Gas Combined Cycle (NGCC) fired power plants, without and with conventional solvent-based CO_2 post combustion capture technology to operate flexibly in response to the demand of the electricity market and factors that may constrain the operation flexibility and cost implications were evaluated by process modelling (IEAGHG 2012).

Scale-up issues related to conventional solvent-based CO_2 post combustion capture technology for Supercritical Pulverized Coal (SCPC) 900MWe and NGCC 809MWe power plant are also evaluated in this work (IEAGHG 2013). However due to the different scope requirement for operational flexibility and scale-up study, the design of solvent based CO_2 post combustion capture process may not be identical for both studies. The difference in the CO_2 capture process may differ mainly in the heat integration, absorber design and solvent characteristics. Although this difference will not change the main outcome on operation flexibility and scale up issues for post combustion capture technology.

2.CO_2 Post Combustion Capture Technology Operational Flexibility

The introduction of post combustion CO_2 capture may impose additional constraints on the start-up and fast load changing of a power plant but techniques are available to overcome these constraints. In an NGCC plant the gas turbine starts up more rapidly than the heat recovery steam generator (HRSG) and the steam turbine. The regenerator in the CO_2 capture plant requires steam from the HRSG or steam turbine and the regenerator needs to be heated to its operating temperature. To avoid constraints on start-up time and to avoid CO_2 emissions during start up, the CO_2 absorber could be operated using lean solvent from a storage tank and the CO_2 rich solvent from the absorber would be stored and fed to the regenerator later. This would enable an NGCC or USCPC plant with CO_2 capture to start up and change load as quickly as a plant without capture. This technique is evaluated in this work is discussed later in this paper.

2.1.Turn off or turn down of CO_2 capture

The net power output of a plant could be increased by turning down or turning off the CO_2 capture and compression units and emitting more CO_2 to the atmosphere. The ability of a plant with CO_2 capture to ramp up power output could in principle be better than that of a plant without capture if the load of the CO_2 capture unit was reduced at the same time as the load of the power generation unit was increased. This work assessed the option of turning off CO_2 capture but various intermediate options involving turning off or turning down parts of the CO_2 capture plant may also be attractive. Turning down or turning off capture would increase emissions of CO_2 to the atmosphere so regulations would have to permit CCS plants to emit more CO_2 during times of peak power demand. Turning down or turning off CO_2 post combustion capture would reduce the plant's internal consumption of electricity and the low pressure steam that would otherwise be consumed by the CO_2 capture unit could be used to further increase the net power output, provided the plant was built with the necessary extra low pressure turbine capacity.

Table 1, Turning off CO_2 capture plant

	NGCC	USCPC
Increase in power output with no capture, %	15.9	27.4
Thermal efficiency, %		
Reference plant with capture	50.6	34.8
Plant with capability to turn off capture	50.2	34.2
Plant with capture turned off	58.6	44.3
Capital cost		
Change in cost per kW of normal output, %	+5.8	+3.9
Change in cost per kW of peak output, %	-8.7	-18.5

Cost of extra peak power capacity, ?/kW	354	322
CO_2 emissions		
Tonnes CO_2 per MWh of extra peak power	2636	2944

It can be noticed from Table 1 that having the capability to turn off capture increases the capital cost of the plant (per kW of normal power output), mainly because of the need for greater steam turbine capacity, but the cost per kW of peak power output is lower. The net capital cost per kW of extra peak power generation capacity is relatively low, probably less than the cost of other types of peak generation capacity.

2.2.Solvent storage

Solvent from post combustion capture can be stored during times of peak power demand for regeneration during times of lower power demand. This reduces the requirement for other peak generation capacity. The extra generation during peak times would have low CO_2 emissions, unlike the alternatives of by-passing CO_2 capture or using peaking plants such as simple cycle gas turbines without CCS.

Table 2, Storage of post combustion CO_2 capture solvent

Power plant type	**NGCC**	**USCPC**	**NGCC**	**USCPC**
Storage scenario	Weekly	Weekly	Daily peak	Daily peak
Hours per week of peak output	80	80	10	10
Increase in power output at peak times, %	6.2	4.8	12.1	22.2
Thermal efficiency, %				
Â Reference plant efficiency, 100% load	50.6	34.8	50.6	34.8
Â Reference plant time weighted average efficiency	50.6	33.6	50.6	33.6
Â Storage plant time weighted average efficiency	45.3	33.5	50.5	33.6
Capital cost				
Â Change in cost per kW of normal output, %	+19.6	+6.1	+9.3	+5.8
Â Change in cost per kW of peak output, %	+12.6	+1.2	-2.6	-13.5
Â Cost of extra peak generation, ?/kW	3116	2891	752	589
Solvent storage				

Â Quantity of solvent storage, 10^3m^3	286	199	30	46

Two operating scenarios described below were assessed in this study as an illustration but it is recognised that in reality power plant operations will depend on many external factors which may change during the operating life of a plant. The ?weekly' and ?daily' scenarios involve different amounts of solvent storage and peak load operation. It can be noticed from Table 2, that the solvent storage has very little effect of the thermal efficiency except for the NGCC weekly scenario, in which one of the gas turbines has to operate at minimum environmental load at off-peak times to regenerate solvent. The solvent storage tanks are conventional sized tanks as used at oil refineries but they are nevertheless large, particularly in the weekly scenario. In the weekly scenario the capital cost per kW of additional peak generation capacity is greater than the cost of the reference power plant, which indicates that this scenario is unlikely to be attractive. In the daily scenario the capital cost per kW of additional peak generation capacity is less than the cost of the reference plant but it is probably higher than the cost of the leading alternative technology for peak load generation, namely simple cycle gas turbines. Solvent storage may be attractive in this scenario, depending on fuel prices, carbon emission costs and the electricity price profile.

3.CO_2 Post Combustion Capture Technology Scale-up Issues

CO_2 post combustion capture design cases for SCPC and NGCC were reviewed to identify the major issues that will likely be faced when moving from the current demonstration scale to constructing and operating large commercial scale units. In general, there appear not to be any major risks which have not been addressed either in power generation or elsewhere in the heavy industrial sector. Integration of the each component of the CO_2 capture facility at large scale and incorporation into existing power plant designs may represent the largest challenge. The main issues related to scale up for post combustion capture process which emerged in the study are construction of absorber and regenerator, stiffing of the boiler, ductwork and flue gas equipment, bypass of flue gas, increase in gas turbine back pressure, amine emission issues etc.

4.References

IEAGHG, Operating Flexibility of Power Plants with CCS, 2012/6, June, 2012.

IEAGHG, Post Combustion CO_2 Capture Scale-up Study, 2013/05, February 2013.

(535a) Modeling and Optimization of a MOF-Based Pressure/Vacuum Swing Adsorption Process Coupled With Material Selection

Gao, H., Northwestern University
You, F., Northwestern University
Snurr, R., Northwestern University
Originally presented on: 11/6/2013 15:15:00 - 15:37:00

The huge amount of carbon dioxide emitted every year has raised significant concern about global climate change. Extensive attention has been paid to Carbon Capture and Sequestration (CCS) to reduce CO_2 emission from large, centralized sources such as power plants. Pressure or vacuum swing adsorption (PSA/VSA) is a promising technology for carbon capture due to its relatively better separation performance, lower energy consumption and lower cost [1]. In order to achieve good performance with PSA/VSA, process optimization and material selection are two important considerations. Process variables, like adsorption and desorption pressures, and material properties, like the isotherm shape and heat of adsorption, can significantly influence process performance. Most previous works have considered these two aspects separately. They either optimize process variables with a fixed material [2, 3], or select materials independent of detailed process modeling [4, 5]. This separated approach has the drawbacks that the performance of a process configuration can be limited by improper choice of adsorbent and that material evaluation without a process model can lead to a non-optimal selection. In this work, we integrate material selection into process modeling by adding integer variables to model material selection. This will enable us to simultaneously choose materials as well as optimize process performance by solving a mixed integer dynamic optimization (MIDO) problem. Our objective is to maximize CO_2 purity and recovery at the same time.

A standard Skarstrom PSA/VSA cycle is modeled with the introduction of integer variables to model the selection of adsorbent materials from a number of promising candidates, such as zeolite 13X, zeolite 5A, ZIF-78, Ni-MOF-74, etc. In a Skarstrom cycle, there is a conflict between purity and recovery. In order to achieve good separation performance, we have two objectives: maximizing CO_2 recovery and purity at the same time. The resulting problem is a partial differential equations (PDEs) constrained multi-objective MIDO program. Multi-objective optimization can be solved using the ε-constraint method, which generates a Pareto-optimal curve by sequentially solving a set of subproblems with different ε values. The governing partial differential equations are discretized completely both in time and space. We apply a finite volume method to discretize the spatial domain and a Radau collocation scheme for temporal discretization. This transforms the MIDO problems to mixed-integer nonlinear programs (MINLP)which can be solved using general-purpose MINLP solvers.

Optimization is performed for several candidate materials used for PSA/VSA purposes. The accuracy of the optimal solutions are verified using MATLAB simulation in which the PDEs are only discretized in space and resulting ODEs are solved using ODE15s in MATLAB. Optimal solutions show the optimal operating conditions and best separation performance for the PSA/VSA cycle. Meanwhile, we identify the best material for the process. The Pareto-curve helps us understand the tradeoff between decision variables under different recovery requirements.

(535b) Energy and Cost Estimation for CO_2 Capture From Dry Flue Gas Using a Vacuum Swing Adsorption Process

Susarla, N., National University of Singapore
Haghpanah, R., Stanford University
Karimi, I. A., National University of Singapore
Farooq, S., National University of Singapore
Rajendran, A., Nanyang Technological University
Originally presented on: 11/6/2013 15:37:00 - 15:59:00

There seems a consensus now that CO_2 emissions to the atmosphere are a major cause for climate change. Nearly 40% of all the CO_2 released annually to the atmosphere is from the fossil-fuel fired power plants. On an average, a 500 MW coal-fired power plant emits approximately 3 million tons of CO_2 per annum [1]. Thus, the capture and concentration of CO_2, especially from power-plant flue gas, is crucial. However, industrially, this remains a challenge because the energy and cost demanded by the capture process is prohibitive. Thus, it is imperative to develop an efficient and economical CO_2 capture technology that can be readily retrofitted into existing power plants.

Recent studies [2-5] suggest that an adsorption-based CO_2 capture can significantly reduce energy penalty as compared to amine-based absorption. Haghpanah et al. [4, 5] presented a comprehensive study of a variety of VSA configurations and their performance for CO_2 capture. However, most studies [2-5] in the literature have used specific energy consumption (kWh per tonne of CO_2 captured) as the sole indicator for comparing various capture methods, and neglected the issues of capital expenditure and plant footprint, which can be equally critical. Recently, Hassan et al. [6] proposed a simplified strategy to estimate the cost of a PSA/VSA process for CO_2 capture using a single column.

In this work, we study and optimize in detail the best 4-step cycle identified by Haghpanah et al. [5] based on Zeochem zeolite 13X as the adsorbent for capturing CO_2 from a dry flue gas. We use the comprehensive model of Haghpanah et al. [4] for simulating this process, which assumes dual-site Langmuir isotherm models and a linear driving force model for the gas-to-solid mass transfer. Owing to the high computational demand of the detailed model, we develop a surrogate model based on the Kriging approximation for the same and use that for optimization. The model based on kriging approximation enables us to do a constrained optimization. In addition, we develop an iterative algorithm that shuffles between the detailed and surrogate models for our single objective optimization. We present a detailed analysis of the various design and operational factors that affect the minimum energy required to capture CO_2 with a purity of 90 mol% purity and a recovery of 90% from a dry flue gas using the proposed adsorption process. Also, we also determine the trade-off between energy, purity, recovery, and productivity for feeds from various emission sources. Furthermore, we examine the advantages of multi-stage separation. Finally, we present a procedure to design and operate a multi-column capture system for dry flue gas from a

500 MW coal-fired power plant and develop a mixed integer non-linear programming (MINLP) formulation based on the kriging approximation model to minimize the total annualized cost. We solve this MINLP model in GAMS using IPOPT and the detailed model in MATLAB to analyse the relative importance of energy consumption, land cost, adsorbent cost, plant footprint, operating expenditure (OPEX), and annualized capital expenditure (CAPEX) for such a capture plant.

References

[1] The future of coal: options for a carbon-constrained world. Massachusetts Institute of technology, Cambridge, MA, 2007.

[2] Dowling AW, Vetukuri SRR, Biegler LT. Large-scale optimization strategies for pressure swing adsorption cycle synthesis. AIChE Journal. 2012; 58.12: 3777-3791.

[3] Zhang J, Webley PA, Xiao P. Effect of process parameters on power requirements of vacuum swing adsorption technology for CO2 capture from flue gas. Energy Conversion and Management. 2008; 49(2), 346-356.

[4] Haghpanah R, Majumder A, Nilam R, Rajendran A, Farooq S, Karimi IA, Amanullah M. Multi-objective Optimization of a 4-step Adsorption Process for Post-combustion CO2 Capture Using Finite Volume Technique. Industrial & Engineering Chemistry Research. 2013; 52(11), 4249-4265.

[5] Haghpanah R, Nilam R, Rajendran A, Farooq S, Karimi IA. Cycle Synthesis and Optimization of a VSA Process for Post-combustion CO2 Capture. AIChE Journal. Under review, 2013.

[6] Hasan MMF, Baliban RC, Elia JA, Floudas CA. Modeling, Simulation, and Optimization of Postcombustion CO2 Capture for Variable Feed Concentration and Flow Rate. 2. Pressure Swing Adsorption and Vacuum Swing Adsorption Processes. Industrial & Engineering Chemistry Research. 2012; 51.48, 15665-15682.

(535c) CO2 Capture From Flue Gas By PSA Utilizing a Structured Adsorbent

Mohammadi, N., University of South Carolina
Abdollahi, A., University of South Carolina
Nicholson, M. A., University of South Carolina
Ebner, A. D., University of South Carolina
Holland, C. E., University of South Carolina
Ritter, J. A., University of South Carolina
Originally presented on: 11/6/2013 15:59:00 - 16:21:00

A cost effective pressure swing adsorption (PSA) process is under development for post-combustion CO_2 capture from coal fired power plants. Several research groups from around the world are showing that PSA is still a viable option, as no separation process has emerged as a clear winner for CO_2 capture. PSA is also attractive because it requires only electricity; it does not require any steam from the power plant. This project is being carried out with funding from DOE NETL and four collaborative partners:

The key challenge to the success of this project is two-fold: 1) an attrition resistant and low pressure drop structured adsorbent must be developed based on commercial zeolite that is compatible with the high velocities associated with shorter PSA cycle times, and 2) a PSA cycle configuration must be developed in concert with the structured adsorbent so that the resulting PSA process delivers an exceptional performance at reduced capital and operating costs. Since the rapid PSA concept is conceptually simple but difficult to implement, a select team has been assembled to carry out this ambitious project and demonstrate the feasibility of the PSA concept for CO_2 capture at the bench scale. This team includes researchers from USC, W. R. Grace, Catacel, and Battelle.

To date, a key development includes the development of a low energy, high feed throughput PSA cycle for CO_2 capture via PSA process simulation. This was an incredible achievement when considering that the bulk density was reduced from 710 kg/m^3 (typical for a packed bed of zeolite beads) to 400 kg/m^3 (entirely possible with a Catacel structured adsorbent core). It boasts around 3,000 L(STP)/hr/kg feed throughput and a separations energy of less than 18 kJ/mol CO_2 captured. This presentation will discuss the latest results obtained from Ritter and his group.

(535d) Post-Combustion CO2 Capture Using Structured Bed Loaded With Functionalized SBA-15 Sorbent

Mittal, N.,
Samanta, A., University of Alberta
Segura, J., Alberta Innovates - Technology Futures
Amiri, S.,
Sarkar, P., Alberta Innovates- Technology Futures
Gupta, R., University of Alberta
Originally presented on: 11/6/2013 16:21:00 - 16:43:00

Post-combustion CO_2 Capture Using Structured Bed Loaded with Functionalized SBA-15 Sorbent

Nikhil Mittal[1], Arunkumar Samanta[1], J. Segura[2], Saeid Amiri[3], Partha Sarkar[2], Rajender Gupta[1,*]

[1]Department of Chemical and Materials Engineering, University of Alberta

9107 - 116 St, ECERF, AB, T6G 2V4, Canada, Email: rajender@ualberta.ca, Tel: 780-492-6861, Fax: 780-49-2881

[2] Environment and Carbon Management Division, Alberta Innovates - Technology Futures, Edmonton, AB, Canada

[3]Wave Control Systems. Inc., Edmonton, Alberta, T6E 0C2, Canada

Abstract:

The conventional packed bed for post-combustion CO_2 capture suffers from significantly high pressure drop whereas the fluidized bed finds problems in sorbent attrition and loss. So, there is a need of more reliable and efficient bed configuration, which can offer a lower pressure drop, fast mass transfer kinetics, high working capacity and improved thermal management. An alternative configuration to packed bed is sorbents placed in structured bed reactor. The major advantage of using structured bed for post-combustion CO_2 capture is low pressure drop because of its straight gas flow channels. The hydrodynamics and external mass transfer can be optimized in this structured bed by adjusting channel size and channel shape, whereas the sorbent loading and/or internal mass transfer can be controlled by the thickness and structure of sorbent channel wall. The objectives of this investigation was to study the adsorption of CO_2 from a flue gas in a structured bed loaded with amine functionalized SBA-15 sorbent and to understand the effect of various design variables such as sorbent bed diameter, channel spacing, membrane layer thickness, and flue gas volumetric flow rate etc.

In the proposed structured bed, the solid sorbent is contained inside a porous alumina tubular membrane reactor (OD ~ 4mm, ID ~ 3mm and average pore diameter ~0.8 micron) and CO_2 laden bulk gas flows through the annular region. CO_2 diffuses across the barrier layer and adsorbs preferentially on the solid sorbent. The equipment was designed and built to host single and multi-tube reactors with different lengths with minimum modification. The concentration of CO_2 stream coming to and from the adsorption column was monitored as a function of time by a Pfeiffer mass spectroscopy (Model: OmniStar GSD 320). Each membrane tube was filled with approximately 0.2 g of 50 wt% PEI impregnated SBA-15 solid sorbent. Thermal swing adsorption cycles were initially tested at different temperatures for CO_2 adsorption using a simulated flue gas containing 10%CO_2 by vol in N_2. The breakthrough curves obtained for CO_2 adsorption from CO_2/N_2 mixture (10%CO_2 bal. N_2) on structured bed of 50 wt% PEI impregnated SBA-15 at 75 Â°C revealed that average CO_2 adsorption capacity was about 2.4 mmol.g^{-1} after ten consecutive cycles. Each adsorption/desorption cycle involved a 10 min of CO_2 adsorption at 75 Â°C followed by regeneration for 15 min at 105 Â°C. For this sorbent, equilibrium CO_2 adsorption capacity measured in TGA under similar condition was 3.07 mmol/g adsorbent. The thermogravimetric data also suggested that the CO_2 capture kinetics was also found to be fast and reached 90% of the total capacities within the first ten minutes. Finally, an optimum set of geometric parameters of the structured bed such as channel spacing and gas superficial velocity has been determined to improve adsorption kinetics and to reduce the pressure drop per unit length of bed.

(535e) A New Temperature Swing Adsorption Configuration for High Purity and Recovery CO2 Capture From Flue Gas With Increased Throughput

Maring, B. J., The University of Melbourne
Webley, P. A., Monash University
Originally presented on: 11/6/2013 16:43:00 - 17:05:00

Tethered amine materials have great potential for CO_2 capture by temperature swing adsorption because of their high CO_2/N_2 selectivity and large heats of adsorption (> 90KJ/mol) which correspond to highly temperature-dependent adsorption isotherms. Tethered amines also have a lower theoretical regeneration energy than aqueous amine solutions because the heat capacity of their solid substrates is lower than that of the water in an amine solution. The major drawback of TSA in industrial applications is the length of time required to heat and cool the large adsorbent beds which can be several hours or even days. We have developed a new TSA configuration which does not rely on external heat transfer and therefore can be run at high throughput, thereby reducing capital costs. Furthermore, the separation can be performed using a narrow temperature range at relatively low temperatures (90ºC to 150ºC), utilizing low quality steam for a reduced energy penalty. We performed numerical simulations of our configuration and achieved high purity CO_2 product with high recovery from flue gas.

(535f) Impact of Water Vapor and Oxygen On Amine Sorbents for Post-Combustion CO2 Capture

Hammache, S., URS-NETL
Hoffman, J. S., U.S. Department of Energy, National Energy Technology Laboratory
Egbebi, A., URS-NETL
Gray, M. L., United States Department of Energy
Originally presented on: 11/6/2013 17:05:00 - 17:27:00

Amine sorbents such as polyethyleneimine/SiO_2 are being considered for post-combustion CO_2 capture from coal-fired power plants. A successful sorbent must be stable under moist conditions present during adsorption and regeneration steps. Also, the presence of oxygen in flue gas may have an impact on the amine sorbents; thus understanding its impact on the amine in the presence of CO_2 is required. In this work, we are evaluating the stability of amine sorbents under steam during regeneration and the impact water and oxygen have on the sorbent during adsorption

The sorbent used in this work was synthesized by wet impregnation of SiO_2 (Fuji Silysia Cariact grade G-10 HPV) with polyethyleneimine (Aldrich PEI Mn-423). CO_2 adsorption-desorption cycling was conducted in a fixed bed reactor. The sorbent was tested prior and subsequent to O_2 pretreatment using dry cycles (10 vol. % CO_2/He at 60°C) to assess any change in its performance. Gas analysis of the reactor effluent was conducted using mass spectrometry.

The impact of 5% O_2/He on the sorbent was investigated at 80 °C. The sorbent was held at these conditions for 5 h, resulting in a decrease in CO_2 capture of 7 mol%. The impact of 5% O_2/He at 80 °C in the presence of water was also investigated. The sorbent was cycled ten times through 5% O_2-8% H_2O/He at 80 °C with the total time adding up to 5 h and regenerated under helium at 80°C. The sorbent was tested under dry conditions prior to and after exposure to O_2/H_2O to assess their impact on its performance. The CO_2 loading was found to decrease by 19 mol% compared to the fresh sorbent. However when the sorbent was exposed to 8% H_2O/He, the decrease in CO_2 loading was found to be only 9 mol %. Clearly the combination of H_2O and O_2 at 80 °C has the greatest negative impact on the sorbent.

The impact of O_2 in the presence of CO_2 on this sorbent and a silanated sorbent will be discussed as well. Characterization of the fresh and spent sorbents prior to and following O_2/H_2O/CO_2/He exposure will be presented.

(535g) Adsorption Isotherm Measurements of Gas Shales for Subsurface Temperature and Pressure Conditions

Wang, B.,
Haghpanah, R., Stanford University
Wilcox, J., Stanford University
Originally presented on: 11/6/2013 17:27:00 - 17:49:00

Carbon dioxide emissions from fossil fuels combustion have been significantly increasing compared to that in the pre-industrial era, which has caused significant increase in global average temperatures. To stabilize atmospheric CO2 emissions, one possible approach may be to inject and store CO_2 into gas shale, where significant amounts of methane are present and can be exploited and recovered. Experimental studies indicate that CO_2 has enhanced adsorption over CH_4, thus the injected CO2 may displace the adsorbed methane inside the gas shale, thereby potentially enhancing methane recovery efficiency. However, the adsorption properties of CO_2 and CH_4 on gas shale are not fully understood, and need to be investigated both experimentally and theoretically.

In our recent work, excess adsorption isotherms of CO_2 on gas shale samples have been measured under subsurface temperature and pressure conditions, using a Rubotherm magnetic suspension balance. The sample used in the study is from the Eagle Ford reservoir located in Texas, United States. Both core chip and powdered forms of sample have been investigated. According to our preliminary results, the shale sample in the powder form always gives higher gas capacity than the same sample in chip form. The ruduced macropore mass-transfer resistance in the powdered sample is a possible explanation for its higher gas capacity.

Since kerogen and clay are the major components contributing to the adsorption behavior in gas shale, adsorption isotherm measurements for carbon-based and clay-based model materials have been performed to determine the relative roles that each of these natural systems play on the overall shale adsorption mechanism and capacity estimates.

In addition, Grand Canonical Monte Carlo (GCMC) simulations have been used to predict the adsorption behavior of carbon-based material with different pore size distributions and chemical heterogeneity. Results from simulation and experiment are compared to further investigate the mechanisms of adsorption in gas shale.

(590a) CFD Simulation of Liquid-Gas Flows On Structured Packing Elements for An Acid Gas Removal Column

Ho, C. S., Hankyong National University
Kim, J. E., Hankyong National University
Nguyen, T. D. B., Hanoi University of Science and Technology
Lim, Y. I., Hankyong National University
Originally presented on: 11/7/2013 8:30:00 - 8:55:00

Structured packing was expected to outperform random/unstructured packings and trays in capacity and/or separation efficiency for a gas/liquid contacting system due to its high surface area and low pressure drop. In this work, computational fluid dynamics (CFD) is applied to study the characteristics of the gas-liquid flows over a structured packing. The interfacial area, liquid hold-up, and pressure drop is examined for the purpose of further investigation on an acid gas removal process in a floating liquid natural gas (FLNG) platform. An Euler-Euler approach combined with the volume-of-fluid (VOF) model is applied to simulate the liquid-gas multiphase flow using a CFD code, Fluent. The standard k-ε model, realizable k-ε, and the renormalization (RGN) k-ε models are compared in the case study. The effects of gas load (F factor) on the liquid film surface and thickness are also analyzed. The numerical results obtained from the simulations are validated with data available in the literature.

(590b) Numerical and Experimental Investigation of Liquid Film Flows On Packings in Absorbers for Post-Combustion CO2 Capture

Iso, Y., IHI Corporation
Huang, J., IHI Corporation
Kato, M., IHI Corporation
Matsuno, S., IHI Corporation
Takano, K., IHI Corporation
Fujiwara, N.,
Originally presented on: 11/7/2013 8:55:00 - 9:20:00

Global warming and greenhouse gas emission reduction have become a major issue in the world. It is necessary to develop Carbon Capture and Storage (CCS) technology for coal fired power plants in order to reduce greenhouse gas emission. According to these backgrounds, Post-Combustion CO_2 Capture (PCC) technology has the great potential as the economical and efficient technology for reducing CO_2 emissions, which can be applied into both retro-fit and new construction of coal fired power plants and can control the CO_2 capture rate.

Gas-liquid interfacial flows, such as the flue gas and liquid solvent, are applied in CO_2 absorbers for PCC in order to capture CO_2 from the flue gas into liquid solvent. Efficient control of these liquid solvent flows by using packing elements in packed columns is important to increase the gas-liquid interfacial area and the mass transfer rate between the gas and the liquid. In typical packed columns, liquid is fed from the top of the packing element, and then the liquid film flow is formed on packing element surfaces. Simultaneously, gas is fed from the bottom of the packing element and makes contact counter-currently flows with the liquid film. There are two types of packing elements: structured packing and random packing. Structured packing elements can increase the gas-liquid interfacial area efficiency and the throughput of gas flow capacity more than random packing elements.

Control of liquid film flows by using packing elements is one of the key design factors in packed columns. In particular, the channeling flow of liquid significantly reduces the gas-liquid interfacial area. To prevent this phenomenon, it is very important to predict the detailed behavior of liquid film flows for the design and development of packing elements. Therefore, the present study focuses on gas-liquid interfacial flows on an inclined wall, which is the model simplified from typical structured packing elements in absorbers for PCC.

There have previously been numerous theoretical, numerical and experimental studies on liquid film flows. These previous efforts have produced useful findings such as the liquid film shape and thickness, liquid hold-up (the amount of liquid held by a packing element), gas pressure drops, validation of Computational Fluid Dynamics (CFD) etc. However, most of these studies concerned liquid film flows on smooth wall surfaces, but the effects of wall surface texture treatments on liquid film flows have not yet been clarified. Furthermore, detailed descriptions of the transition phenomena between the film flow and the rivulet flow, as well as how such phenomena are affected by wall surface texture treatments, are still lacking.

In what follows, this study develops a three-dimensional numerical simulation technique using CFD with the Volume of Fluid (VOF) model as well as a lab-scale experimental testing technique. We investigate that the effects of wall surface texture treatments on the interfacial flow through the comparison of two geometry cases (smooth wall and wavy wall). Furthermore, our fundamental technics are applied to develop our advanced design of surface texture treatments and packing geometry. The absorption column tests using the CO_2-NaOH system is carried out to measure the performance of the absorption rate and the gas pressure drop. Our new findings in this paper include the followings.

(1) With a typical texture geometry used in a commercial industrial packing element, the present numerical and experimental results comparing the two geometry cases (smooth wall and wavy wall) show that the surface texture treatments can help to prevent liquid channeling and can increase the wetted area quantitatively. The main reason for the increase of the wetted area on the wavy wall is that the liquid film break-up is inhibited due to the spreading of the liquid flow in spanwise direction by the wavy wall geometry.

(2) The numerical simulation results, on both of smooth and wavy walls, agree well with the experimental results at points such as the gas-liquid interfacial surface shape and the wetted area. These validations demonstrate that the present simulation, which uses the VOF model, is capable of predicting gas-liquid interfacial flows on an inclined plate with a high accuracy in a reasonable calculation time period.

(3) Our advanced design of surface texture treatments and packing geometry is developed by using present numerical and experimental techniques. The absorption column tests using the CO_2-NaOH system show that our advanced packings trail-manufactured with two types of specified surface areas (300 m2/m3 and 200 m2/m3) have higher absorption performance and lower gas pressure drop by comparing to several conventional structured packings.

In an ongoing project, IHI has constructed a post-combustion capture pilot plant at its Aioi Works (capacity: 20 tons of CO_2/day). This pilot plant can be operated in combination with the co-located coal combustion test facility, which makes it possible to carry out operational evaluations using actual flue gas from coal-fired boiler. Sequential test operations were started at this pilot plant from June, 2012. On the next step, our advanced packing systems will be evaluated by using this pilot plant in 2013.

(590c) Advanced Characterization of CO2 Absorption Into Dispersed Aqueous Amine Solvents

Tamhankar, Y., Oklahoma State University
Whiteley, J. R., Oklahoma State University
Resetarits, M., Fractionation Research, Inc.
Aichele, C. P., Oklahoma State University
Originally presented on: 11/7/2013 9:20:00 - 9:45:00

Sprays have numerous industrial applications in the energy, chemical, pharmaceutical, and agricultural industries. Simplicity of construction and low gas phase pressure drop make spray columns ideal for a variety of applications in the process engineering industry including flue gas desulphurization, gas treating, humidification, and heat transfer. For natural gas treating and post-combustion carbon capture applications, the low gas phase pressure drop can potentially result in significant cost savings. Despite the wide use of sprays, mass transfer in spray systems for CO_2 removal is poorly understood, and design practices for spray columns in these cases are typically empirical. There is a scarcity of reliable absorption data which is essential in developing more robust designs. This paper presents the efforts underway at Oklahoma State University in measuring absorption rates for a variety of CO_2-amine systems in a laboratory spray column. The experimental setup, operating procedures, and measurement techniques will be discussed. Results from the experiments will assess the effect of solvent concentration, gas flow rate, liquid flow rate, gas composition, and nozzle type on the mass transfer. A phase doppler interferometer (PDI) is integrated into the unique experimental setup to measure the droplet size and velocity of the resulting droplets, thereby leading to the measurement of the effective surface area of the spray. The data and methodology presented will be essential for subsequent design of large-scale absorption systems.

(590d) Consistent Measurements of Hydraulic and Mass Transfer Packing Characteristics

Perry, M., University of Texas at Austin
Wang, C., University of Texas at Austin
Seibert, F., University of Texas at Austin
Originally presented on: 11/7/2013 9:45:00 - 10:10:00

Packed columns are generally regarded as the best design for post-combustion carbon dioxide absorbers. These towers will handle substantial gas loads, and as a result, the energy associated with moving the post-combustion exhaust gases through the stacks will be significant. It will also be important to optimize the mass transfer taking place within the packed absorber to minimize energy losses in carbon capture systems. This research utilizes a consistent approach to measuring mass transfer and hydraulic characteristics of both random and structured packings in order to create new and improved models for pressure drop, liquid holdup, contact area, and liquid and gas mass transfer coefficients, thereby improving packed absorber designs. Experimental methods have been created to measure the following characteristics of any packing:

-Pressure drop (dry and irrigated)

-Liquid Holdup

-Flood Point

-Effective Mass Transfer Area

-Liquid Film Mass Transfer Coefficients

-Gas Film Mass Transfer Coefficients

The purpose of this paper is to update the progress of this research by adding the results of characterizations completed on several new packings. This paper will also show progress made in modeling efforts utilizing this new data. Data will also be compared to existing models for contact area, liquid film, and gas film mass transfer coefficients. Updating the previous version of this presentation, 5 new packings have been studied and modeled. This new data will be combined with previous data to create a database of 14 packings.

(590e) Packing Economy Analysis for Post-Combustion CO2 Capture

Wang, C., University of Texas at Austin
Perry, M., University of Texas at Austin
Seibert, F., University of Texas at Austin
Rochelle, G. T., The University of Texas at Austin
Originally presented on: 11/7/2013 10:10:00 - 10:35:00

Packing Economy Analysis for post-combustion CO_2 capture

by Chao Wang[1,2], Micah Perry[2], Frank Seibert[2], Gary T. Rochelle[1]

[1]Department of Chemical Engineering, The University of Texas at Austin

[2]Separations Research Program, The University of Texas at Austin

Austin, TX 78712

Packing is widely used in post-combustion CO_2 capture because of its low pressure drop, good mass transfer efficiency, and ease of installation. In the post combustion CO_2 capture process, absorber and stripper costs can be divided into two parts: the capital costs and the operation costs. Economy analysis of these costs has been done by changing the operating gas velocity, liquid flow rate, column diameter, packed height, packing type(including the total area and the corrugation angle) to find the optimum costs.

Capital costs include the one-time investment of building the column, packing costs and auxiliaries such as pump and blower costs. The column and packing costs are determined by the diameter of the column, the packed height, as well as the packing type. The blower purchased costs is determined by the maximum gas flow rate and pressure drop while the pump purchased costs is determined by the maximum liquid flow rate. Operation costs include the running costs and the maintenance costs. The running costs are related to the work of the blower and pump while the maintenance costs are estimated.

The column is operated with 7m MEA at 40Â°Ã¦ with a capacity of approximately 1 mol CO_2/kg solvent. The required Number of Transfer Unit (NTU) is calculated based on 90% removal of CO_2 with an inlet CO_2 concentration of 12mol%. The Height of Transfer Unit (HTU) is calculated by equation $HTU=u_G/K_{OG}a_eRT$. The effective area is calculated from modified Tsai's area model:

The overall mass transfer coefficient K_{OG} is calculated from the gas film mass transfer coefficient k_G, liquid film mass transfer coefficient k_L, and the mass transfer coefficient related to the chemical reactions by this equation: $\frac{1}{K_G} = \frac{1}{k_G} + \frac{H_{CO2}}{E^* k_L} + \frac{1}{k_L}\frac{\partial P^*_{CO2}}{\partial [CO_2]}$

The chemical reaction constant can be obtained from reports; k_G and k_L values for different packings are from mass transfer models developed from this work based on measured values. Thus, the packed height can be calculated.

For this case, the column is operated with a fixed ratio of gas and liquid flow rate (L/V). The column diameter is calculated by the operating gas velocity (u_G). Therefore, the variables can be reduced to two: the gas velocity and the packing type. Eight structured packings with different total area and corrugation angle are chosen to run case by case. An Excel program including all the required information to calculate the total costs is build. The SOLVE function is utilized to optimized the total costs by varying gas velocity throughout the operating range.

For data analysis, the equipment costs are converted to total fixed plant costs by a multiplication factor of 4 based on the Chilton method. The total fixed plant costs is then converted to annualized costs with the assumption of a 1-year start-up time, a 2-year construction time, and a 10-year project life. Finally, the overall costs are described as the costs per tonne CO_2 by dividing the annualized costs and the annual CO_2 output of the plant.

Results of economy analysis for each packing are shown. For each case, the optimized gas velocity, column diameter, packed height, and the minimized costs per tonne CO_2 are listed in tables. The figure of overall costs varying with operating gas velocity is shown. Figure 1 shows the case for packing MP250Y, running in a pilot plant with an annual output of 250MW. Comparison of minimum costs between different packings is made to find the optimized packing total area and corrugation angle.

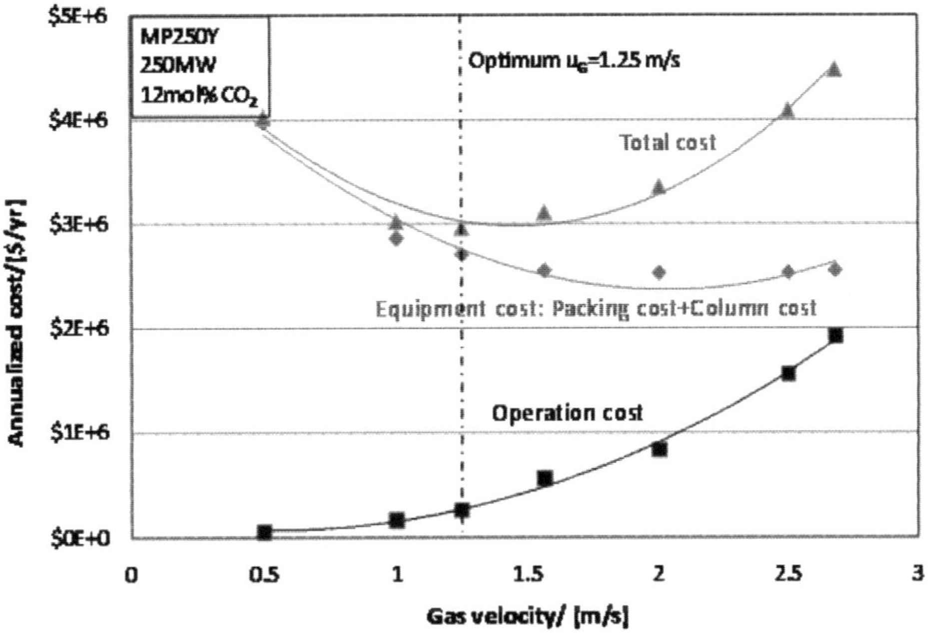

Figure 1. Economy analysis for MP250Y in a 250MW pilot plant

(590f) Investigating the Impact of Non-Isothermal Behavior and the Role of Water Transfer in Membrane Contactors for CO2 Capture

Neveux, T., EDF R&D
Le Moullec, Y., EDF R&D
Castel, C., Laboratoire RÃ©action et GÃ©nie des ProcÃ©dÃ©s LRGP- CNRS, UniversitÃ© de Lorraine
Favre, E., Laboratoire RÃ©action et GÃ©nie des ProcÃ©dÃ©s LRGP- CNRS, UniversitÃ© de Lorraine
Originally presented on: 11/7/2013 10:35:00 - 11:00:00

The capture of carbon dioxide from industrial sources is an attractive option to reduce the atmospheric CO_2emissions. Among the different separation techniques, chemical absorption in amine solvents is expected to be the dominant technology at the early stage of the carbon capture development. The main bottleneck to an industrial deployment lies in the high energy consumption of such processes and subsequently the high cost of capture. Membranes contactors are often foreseen as an efficient way to intensify the gas-liquid absorption process. The idea is to use a porous hydrophobic membrane as a gas-liquid contactor, providing a large and controlled interfacial area between the gas and liquid phases. Such a contactor also allows the use of volatile solvent such as ammonia or DMEA instead of classical amines. Obviously, the addition of a supplementary medium compared to classic gas-liquid contactors such as packing induces an increased mass transfer resistance. Still, the use of membrane contactors provides better overall transfer performances than packing due to the (4 to 20 times) higher interfacial area and allows to operate in a much wider range of phase velocity.

Performances of membrane contactors are widely studied in literature both from experimental and modeling points of view but studies on mass transfer performance almost systematically postulate isothermal conditions and dry flue gases for sake of simplification [1]. However, the water fraction of flue gases is not negligible since flue gases are saturated in water during depollution (e.g. about 10% for flue gases from coal combustion). Moreover, the process is not isothermal in industrial conditions due to the water condensation and evaporation and to the high heat of reaction between $CO_2$and amine.

Consequently, there is clearly a room for comprehensive analysis of these effects in representative conditions and this work intends to highlight the impact of non-isothermal behavior and the role of water transfer on CO_2capture performance in membrane contactors.

In order to do so, numerical simulations have been performed with a rigorous phenomenological model of chemically enhanced heat and mass transfer in membrane contactors. Thermodynamics of electrolyte solutions (H_2O-CO_2-amines) is modeled with the extended UNIQUAC model, particularly well adapted for the prediction of vapor-liquid and chemical equilibria in amine solution. A rate-based formulation is used for the modeling of heat and mass transfer with consideration of the chemical enhancement due to reaction between CO_2 and amines. The adopted model has already proved its accuracy and ability to represent absorption and stripping operations with packing as gas-liquid contactors [2] and has been adapted to deal with the specificity of the membrane.

Before investigating the desired effect, the model has been validated on available experimental data in thermostated conditions with a dry inlet flue gas. Predicted CO_2 capture efficiency are in excellent agreement with experimental measurements for two membrane materials (PDMS and PP) and in a wide range of gas velocities. Thereafter, the model has been used to predict the CO_2 capture efficiencies in non-isothermal conditions with a saturated gas. The temperature profiles of gas and liquid phases are specifically investigated. From the simulation results, it appears that the presence of water in flue gas significantly influences the capture efficiency. For instance with a representative test, the predicted capture efficiency is in the order of magnitude of 90% if the system is considered isothermal whereas the capture efficiency decreases down to 75% if a non-isothermal behavior is considered. This reduction can be interpreted by observing the temperature profiles and direction of transferred water flux along the contactor. Both gas and liquid temperatures reach a maximum corresponding to an elevation around 30°C with respect the phase inlet. As the liquid flows, its temperature increases due to the heat of reaction and the massive condensation of water from flue gas. Although the temperature rise favors the kinetic between CO_2 and solvent, the significant shift of gas-liquid equilibrium (reduction of CO_2 solubility) explains the decrease of capture efficiency. The temperature decreases as the liquid gets closer to the gas inlet due to water evaporation and conductive heat transfer with the gas phase.

In this study, insights are given on the impact of water in flue gases and the consideration of non-isothermal effects on CO_2 capture efficiency through, using rigorous phenomenological modeling. The consideration of water transfer due to condensation and evaporation as well as heat of reaction leads to a reduction of predicted capture efficiencies. This more representative approach will allow to predict more accurately the intensification factor which can be achieved by the use of membrane contactors instead of classical gas-liquid contactors.

[1] Favre E., Svendsen H.F. (2012). Membrane contactors for intensified post-combustion carbon dioxide capture by gas-liquid absorption processes. Journal of Membrane Science,407-408, 1-7.

[2] Neveux T., Le Moullec Y., Corriou, J.P., Favre, E. (2013). Modeling CO_2 capture in amine solvents: Prediction of performances and insights on limiting phenomena. Industrial & Engineering Chemistry Research, 52(11), 4266-4279.

(640a) The Influence of Plasticization and Dual-Mode Competetive Effects On CO2 Transport in Thermally Rearranged (TR) Polymers

Gleason, K. L., The University of Texas at Austin
Smith, Z. P., The University of Texas at Austin
Liu, Q., The University of Texas at Austin
Paul, D. R., The University of Texas at Austin
Freeman, B. D., The University of Texas at Austin
Originally presented on: 11/7/2013 8:30:00 - 8:55:00

In this work, results of mixed-gas permeation tests involving CO_2 permeation in TR polymers derived from 3,3'-dihydroxy-4,4'-diamino-biphenyl (HAB) and 2,2'-bis-(3,4-dicarboxyphenyl) Hexafluoropropane Dianhydride (6FDA) are presented. Polyimide HAB-6FDA films were converted to polybenzoxazole (PBO) to varying degrees by following different thermal treatment protocols. Permeation tests with these films were conducted using CO_2-containing mixtures at temperatures from -15 to 70°C, and pressures up to 50 atm. The bulk of the results are for 50% CO_2/CH_4 mixtures. Preliminary results for CO_2/N_2 mixtures are presented as well, and exhibit trends similar to the CO_2/CH_4 results. Mixed-gas results are compared against pure-gas measurements. CO_2/CH_4 mixed-gas selectivity at 35°C was 20-30% higher than the simple ratio of pure-gas permeabilities. Strong dual-mode competitive effects were responsible for a decrease in CH_4 and N_2 permeability in the presence of CO_2, with very little change to CO_2 permeability in the mixture. Plasticization is not evident in the pure-gas CO_2 permeation data. However, in the mixed-gas experiments, CH_4 permeability was found to increase gradually with increasing CO_2 pressure. The degree of plasticization was found to decrease as the degree of thermal conversion to PBO increased.

(640b) Facilitated Transport in Hydroxide-Exchange Membranes for Post-Combustion CO2 Separation

Xiong, L., University of Delaware
Gu, S., University of Delaware
Jensen, K., University of Delaware
Yan, Y., University of Delaware
Originally presented on: 11/7/2013 8:55:00 - 9:20:00

Facilitated transport, hydroxide-exchange membranes have been fabricated for the separation of CO_2 from post-combustion flue gas feed. First, a correlation between expected basicity of the fixed-site functional group and CO_2 separation performance is highlighted. Later, this observation is used to suggest that phosphonium may be a good candidate for achieving high separation performance. Separation tests show that quaternary phosphonium-based membranes achieve separation performance above Robeson's Upper Bound, an empirical bound describing the inverse relationship between permeability and selectivity. Specifically, permeabilities as high as 1090 and selectivities as high as 275 have been achieved. As expected, increases in CO_2 permeability are observed when increasing the degree of functionalization. In contrast, there is an small decrease in CO_2/N_2 selectivity when increasing the degree of functionalization, attributed to the increasingly large water uptakes at higher degree of functionalization. The high performance found in the membrane can be attributed to the combination of the quaternary phosphonium moiety and efficient transport channels.

(640c) CO2 Removal From High Pressure Natural Gas With Hybrid Fixed-Site-Carrier Membranes: Experimental, Process Design and Feasibility Analysis

He, X., Norwegian University of Science and Technology
Kim, T. J., SINTEF
Hagg, M. B., NTNU
Originally presented on: 11/7/2013 9:20:00 - 9:45:00

Natural gas (NG) is becoming one of the most attractive growing fuels for world primary energy consumption. However, the raw natural gas usually contains considerable amount of CO_2 which should be removed to meet the natural gas network grid specifications. Membrane systems possess of small footprint, low capital and operating costs, being environmentally friendly, and exhibiting process flexibility, which show a great potential for CO_2 removal from natural gas even though it has only 5% of the market today. Commercial gas separation membranes for natural gas sweetening are mostly made from cellulose acetate (CA), cellulose triacetate (CTA) and polyimide (PI). The challenges of relatively low separation performance (both CO_2 permeance and CO_2/CH_4 selectivity) and membrane plasticization for these commercial polymeric membranes direct to the development of novel, high performance composite membranes. Our previous work has already documented that the carbon nanotubes (CNTs) reinforced PVAm / polyvinyl alcohol (PVA) blend FSC membranes could improve the membrane performance, especially at high pressure operation [1]. Thus, the high performance hybrid FSC membranes were prepared using an optimized preparation condition in this work.

Carbon nanotubes reinforced PVAm/PVA blend FSC membranes were prepared to improve the mechanical strength and reduce the compaction influence at high pressure operation. A special casting procedure was employed to coat an CO_2-selective layer with a pre-determined thickness as reported by Kim et al. [2]. An optimized heat treatment procedure is conducted to enhance the membrane mechanical strength, separation performance and stability. The size of the prepared large flat-sheet hybrid FSC membrane is 30 cm Ã—30 cm.

Membrane separation performance was characterized by high pressure gas permeation testing with the synthetic mixed gases of 10% CO_2 - 90% CH_4 and 50% CO_2 - 50% CH_4. The high pressure pilot-scale module involves 3 sheets (the membrane area of each sheet is 110 cm^2) that can be operated in parallel or individually. A constant feed flow of 3000 Nml/min was humidified (RH 100%) and fed into the membrane module at 30 Â°C and different pressures without sweep gas, and the developed FSC membranes showed a high CO_2 permeance of 0.084 - 0.218 m^3 (STP) / (m^2.h.bar) with a CO_2/ CH_4 selectivity of 17.9 - 34.7 at different feed pressures from 40 to 10 bar. Moreover, the membrane capacity was also investigated by testing the whole module (area: 330 cm^2) at 30 Â°C and 30 bar using different feed flow rates. It was found that the CH_4 purity in retentate can achieve the requirement (i.e., >96%) using a single stage membrane unit at a low feed flow rate (high stage-cut), but the CH_4 losses are quite high which indicated that part of methane should be recovered using a 2nd?stage membrane unit- this could be used to guide the process simulation.

Based on the experimental data, process simulation was conducted by HYSYS integrated with an in-house membrane program (ChemBrane) to evaluate the process feasibility of membrane systems for CO_2 removal from high pressure natural gas. A two-stage membrane system was designed for process feasibility analysis. Process simulation was conducted on the basis of a natural gas feed flow of 5Ã—10^5 Nm^3/h contains 10% CO_2 / 90% CH_4. The minimization of specific power consumption and natural gas sweetening cost (economic cost estimation using CAPCOST_2008 [3]) were applied to process optimization, and the results are shown in Table 1. It can be found that membrane process with the developed FSC membranes was feasible for CO_2 removal, even at a relatively low CO_2 feed concentration (~10 %) in natural gas, and it is possible to achieve > 96 % CH_4 purity in the retentate stream and a low CH_4 loss < 2%. A minimum cost of 5.91e-3 \$/ Nm^3 sweet natural gas produced was found at a feed pressure of 10bar in the 2nd stage to achieve the given separation requirements (i.e., CH_4 purity > 96%, CH_4 losses < 2% and captured CO_2 purity > 95%), which is lower compared to 6.40e-3 \$/ Nm^3 product with amine absorption [4]. Thus, the developed CNTs reinforced hybrid FSC membranes show great potentials for CO_2 removal from high pressure natural gas.

Table 1 Simulation results

Parameters	Simulation results
Feed pressure: 1st / 2nd stage , bar	40/10
Sweet NG productivity, Nm^3/h	4.67E+05
CH_4 purity in sweet NG, %	96.00
CH_4 losses, %	0.35
CO_2 purity, %	95.19
CO_2 recovery, %	62.50
Total membrane area, m^2	2.62E+05
Specific power consumption, kWh/Nm^3 sweet NG	4.11E-02
NG sweetening cost, \$/Nm^3 sweet NG	5.91E-03

Acknowledgements

Â The authors acknowledge the NaGaMa project (partners: Norwegian Research Council, Statoil and Petrobras) for financing this work. The high pressure pilot-scale module designed by PHILOS in South Korea is also acknowledged. The authors also thank SHOWA DENKO K. K. company in Japan to provide the carbon nanotubes for this study.

References

[1] He
X, Hägg MB. Hybrid Fixed-site-carrier Membranes for CO_2/CH_4 Separation. Euromembrane 2012. London, UK.

[2] Kim
T-J, Vrålstad H, Sandru M, Hägg M-B. Separation performance of PVAm composite membrane for CO2 capture at various pH levels. J Membr Sci. 2013;428(0):218-24.

[3] Turton
R, Bailie RC, Whiting WB, Shaeiwitz JA. Analysis, Synthesis, and Design of Chemical Processes, Third Edition. Prentice Hall, New Jersey. 2008.

[4] Peters
L, Hussain A, Follmann M, Melin T, Hägg MB. CO2 removal from natural gas by employing amine absorption and membrane technology?A technical and economical analysis. Chem Eng J. 2011;172(2?3):952-60.

ADDIN EN.REFLIST

(640d) Synthesis, Characterization, and Gas Transport Properties of Two Isomeric Thermally Rearranged (TR) Polymers for Gas Separation Membranes

Liu, Q., The University of Texas at Austin
Borjigin, H., Virginia Tech
Paul, D. R., The University of Texas at Austin
McGrath, J. E., Virginia Tech
Freeman, B. D., The University of Texas at Austin
Originally presented on: 11/7/2013 9:45:00 - 10:10:00

Thermally rearranged (TR) polymers, which are typically polybenzoxazoles with a rigid-rod structure, have recently gained interest in gas separation application due to their excellent gas transport and separation performance, particularly for CO_2/CH_4 mixtures. In fact, for these separations, TR polymers have permeabilities and selectivities that are above the Robeson upper bound. Another attractive feature is their apparent strong resistance to plasticization. In this study, the permeabilities of H_2, N_2, O_2, CH_4, and CO_2 were measured over pressure ranging from 3 atm to 17 atm at 35°C for aromatic polyimides based on 3,3'-diamino-4,4'-dihydroxy biphenyl and 2,2'-bis-(3,4-decarboxyphenyl) hexafluoropropane dianhydride (AHB-6FDA). Thermally rearranged (TR) polybenzoxazoles (PBO) were obtained by thermally treating the polyimides between 350 and 450°C. The the gas transport properties of AHB-6FDA are compared with those of its isomer based on 3,3-dihydroxy-4,4'diamino-biphenyl and 2,2'-bis-(3,4-decarboxyphenyl) hexafluoropropane dianhydride (HAB-6FDA). Both polyimides and the corresponding TR polymers showed similar gas transport properties.

The effects of polyimide precursor ortho-positioned functional groups on gas transport properties were also investigated. Acetate-containing AHB-6FDA displayed much higher gas permeability and comparable selectivity relative to hydroxyl-containing AHB-6FDA for both the polyimide and its corresponding TR polymers. Thus adding bulkier groups at the ortho-position improves gas transport performance in these polymers.

(640e) Process Modeling and Optimization of Post-Combustion Carbon Capture With N2-Selective Membranes

Narakornpijit, K., Stanford University
Yuan, M., Stanford University
Haghpanah, R., Stanford University
Wilcox, J., Stanford University
Originally presented on: 11/7/2013 10:10:00 - 10:35:00

Membrane separation has been proposed as a post-combustion carbon capture technology for its potential advantages such as a small footprint, environmental friendliness (i.e., no solvent involved in the process), and a lower energy requirement. Current modeling and optimization work has focused on post-combustion capture using polymeric, CO_2-selective membranes, where the CO_2 in flue gas is selectively permeated through the membrane and is concentrated in the permeate stream. One of the limitations of CO_2-selective membranes is that the low CO_2 concentration in flue gas (12–15 vol% from coal-fired power plants, 5–8 vol % from natural gas–fired power plants) provides a low driving force for separation, which requires that multistage separation and/or stream recycling be employed to achieve DOE's capture target of 90% capture (removal efficiency) and 95% product purity.

The concept of a N_2-selective membrane has been explored as a post-combustion carbon capture technology with the goal of further reducing the land and energy requirements of membrane-based capture processes. Two factors may contribute to the potential energy savings. First, the predominantly higher N_2 concentration in flue gas may provide a higher driving force for membrane separation, eliminating part of the feed compression requirements. Second, the concentrated CO_2 stream exits from the high-pressure retentate stream, which may displace part of the compression energy required for pipeline transport. The concept of N_2-selective membranes has been experimentally proven with pure vanadium (V) and its alloys with ruthenium (Ru) as membrane materials. In our modeling work, flue gas from a coal-fired power plant containing ~15 vol% CO_2 is treated with an ideal N_2-selective membrane (assuming minimal defects present in the material, and only N_2 passes across the membrane) in the countercurrent flow scheme. The membrane optimization literature mostly deals with single-objective optimization of costs or energy use. However, a capture plant should satisfy two criteria that have direct implications for operating and fixed costs, namely low parasitic energy consumption and a small footprint. In the present study, we performed multi-objective optimization using Genetic Algorithm (GA) to minimize energy consumption and membrane area simultaneously. We also conducted parametric analysis to investigate the potential of hybrid N_2- and CO_2-selecitve membrane configurations to further lower the energy consumption of the capture processes. The results of these studies will be presented.

(640f) Benchmarking Of Hydrogen Selective Membranes Through Experiments and Modelling

Pieterse, J. A. Z., Energy Research Centre of the Netherlands
Boon, J.,
Dijkstra, J. W., Energy research Centre of the Netherlands
Originally presented on: 11/7/2013 10:35:00 - 11:00:00

Palladium (-alloy) membranes are a promising option for hydrogen separation in industrial hydrogen production and in pre-combustion CO_2 capture [1,2]. Given the state of the art, benchmarking of membranes from different vendors is a crucial step in the demonstration of the maturity and performance under industrially relevant conditions. A multitude of membrane manufacturers and developers exists, yet at the moment, performance figures have been published at different conditions, and including the impact of externals, e.g. different membrane module designs. The present contribution provides an approach to assess the performance of H_2-selective membranes under industrially relevant conditions, independent from the impact of module design. The strategy presented here, consisting of experiments and model development, will be a valuable tool in benchmarking membrane performance. In order to evaluate membrane performance without artifacts induced by the experimental module, a membrane model has been devised. Systematic experimental tests were performed (extended for up to 1500 hours on stream) with membranes that were delivered by a number of different suppliers, allowing for model development and experimental validation. The staged experimental approach allows for investigating systematically the effects of intrinsic permeation, hydrodynamics, and concentration polarization. Membranes are tested at a scale of 300 to 800 cm^2 in a pilot setup [3] capable of accommodating up to 6 membranes of 100 cm length. Experiments have been performed that started with pure H_2 feed, without sweep, subsequently followed by introducing N_2 on the feed side, and N_2 sweep gas. Using a phenomenological description for the palladium layer and the dusty gas model for the membrane support, coupled to a 2D Navier–Stokes solver with a convection-diffusion equation to account for possible concentration polarisation, all relevant mass transfer resistances are adequately modeled [4]. For the conditions investigated, the main resistances to mass transfer are concentration polarisation in the retentate, hydrogen permeation through the metallic palladium layer, and a diffusional resistance in the support layer. Finally, the effect of inhibition by syngas components was measured for representative gas compositions and incorporated in the permeation equation. The 2D model was then used for evaluating and comparing the intrinsic membrane performance and for predicting membrane performance at commercial scale. In parallel, membrane and membrane-module robustness data extracted from the long term stability experimental testing program provided valuable insight in the technological status.

References

[1]. N.W. Ockwig and T.M. Nenoff, Chemical Reviews 107 (2007) 4078-4110.

[2]. G.Q. Lu, J.C Diniz da Costa, M. Duke, S. Giessler, R. Socolow, R.H. Williams, T. Kreutz, Journal of Colloid and Interface Science 314 (2007) 589.

[3]. D. Jansen, J.W. Dijkstra, Energy Procedia, 1 (2009) 253.

[4]. J. Boon, J.A.Z. Pieterse, J.W. Dijkstra, M. van Sint Annaland, Int. J. of Greenhouse Gas Control 11 (2012) 122.

(647a) The Novel Cal-C Process for Single-Stage Removal of CO2, SOx and NOx – Experimental Studies and Process Analysis and Integration

Deshpande, N., The Ohio State University

Phalak, N., The Ohio State University

Wang, W., The Ohio State University

Fan, L. S., The Ohio State University

Originally presented on: 11/7/2013 8:30:00 - 8:55:00

A multiple-gas cleanup process is proposed for the removal of CO_2, SO_x and NO_x from coal combustion flue gas. This cleanup occurs at high temperatures (500-650 °C) in a single-stage gas-solid fluidized-bed reactor. This process ? named Cal-C process ? utilizes limestone-derived calcium sorbent and inexpensive coal char, and consists of the three-step calcium looping developed at OSU with the addition of char to the carbonator. The calcium sorbent is re-circulated through the system in a closed loop after regeneration and reactivation, while fresh addition of char is warranted due to complete consumption.

The process has been verified experimentally using a lab-scale fluidized-bed reactor setup and the simultaneous removal of all the species of interest has been established. Based on previous studies, lignite coal chars are known to be suitable for this application. In this work, experiments are performed to quantify the consumption of char on the basis of selectivity of char toward NO. Different parameters such as calcium to char loading, O_2 concentration and NO concentration are studied for their effect on this selectivity. Further, the fate of gaseous species such as CO and H_2, which are likely to be produced in-situ, is investigated and possible process conditions are identified for minimizing unwanted byproducts. Different process configurations with respect to the entry of char in the process loop are also analyzed experimentally. Thermogravimetric analysis of the hydration (reactivation) reaction is also conducted to study the effect of presence of char on the hydration kinetics. Using ASPEN simulations, the process analysis is conducted and different process configurations are evaluated with respect to the opportunities for heat recovery, minimization of excess char requirement, etc.

(647b) Carbon Dioxide Absorption Into Aqueous Solution of Hydrophilic Amine-Functionalized Ionic Liquids

Lu, B., Zhejiang University
Wang, X., Zhejiang University
Jin, J.,
Li, W.,
Originally presented on: 11/7/2013 8:55:00 - 9:15:00

Carbon dioxide (CO_2), as one of the greenhouse gases (GHG), is currently responsible for over 60% of the enhanced greenhouse effect. Since global warming was aggravated, carbon capture and sequestration to reduce the release of $CO_2$to atmosphere is of critical importance. Chemical solvents (such as amines) have been extensive used, but still suffer some disadvantages, which restricting the promotion of the technology.

In recent years, Ionic liquids (ILs) are frequently used as solvents for gas absorption operations due to their immeasurably low vapor pressure, which can improve the process economy and general efficiency of gas separations. However, gas dissolution in ILs is a physical phenomenon with no chemical reaction and the absorption rate is quite slow. In order to introduce chemical reaction, a variety of ILs were integrated with ions incorporating functional groups. In the open literature, functional ILs were mostly dissolved in conventional ILs and owns high viscosity, which is unbeneficial to industrial applications.

Herein, we designed ether functionalized imidazolium based ILs with glycine anions for the absorption of CO_2. And a Hydroxy group was also introduced to this ILs in order to improve the solubility of the ILs in aqueous solution. The absorption of CO_2 by hydrophilic amine-functionalized ILs ([C_2OHmim]Gly) and Sodium glycinate (SG) were investigated respectively in a double stirred-cell absorber with a planar gas–liquid interface at 15 kPa CO_2 partial pressure. The absorption rate and capacity of [C_2OHmim]Gly into aqueous solution were higher than SG solution. Moreover, the IR spectra of [C_2OHmim]Gly was also proved that the functionalized ILs containing hydroxyl and glycine anions.

It has been widely confirmed that, vaporization and decomposition during regeneration is one of most non-negligible dissipation of aqueous amine, which increases the capital cost of the process. The regenerate efficiency of solvent is important for industrial application. Therefore, the regenerate experiments were investigated in an oil bath at 375.15 ± 0.5 K, and sustained approximately 90 minutes. The multi-cycle regenerate efficiency of [C_2OHmim]Gly and SG were compared with each other. The regenerate efficiency were all decreased as the regeneration cycle was increased but [C_2OHmim]Gly were decreased quite slow. After the fourth regeneration, the efficiency of [C_2OHmim]Gly just decreased 13%, SG almost decreased

25%. The result indirectly proved that the [C_2OHmim]Gly were well recycled and has a higher stability than SG by thermal regeneration even after multi-cycle experiments

As oxygen is one of the most typical components in the flue gas and may give rise to undesired side reactions, effect of O_2 was investigated and the O_2 concentration gradually increased in a range of 0-8 % (V/V). The absorption rate and capacity of CO_2 into [C_2OHmim]Gly and SG were all decreased in the presence of oxygen, but the decline trend of [C_2OHmim]Gly was relatively small. The capacity of [C_2OHmim]Gly only reduced 5.9 %, but the capacity of SG almost reduced 17 % in 4%O_2 compared with the capacity in absence of O_2. Oxygen is susceptible to induce oxidative degradation of alkanolamines. This undesirable breakdown not only reduces their CO_2 absorption capacity but also introduces unwanted degradation products, thus forcing the solution to be eventually discarded. Therefore, the multi-cycle regenerate experiments in present of oxygen (4%O_2) were also investigated. And the differences of regenerate efficiency of these two solutions were significant.

The above experiment results indicated that the hydrophilic amine-functionalized ILs ([C_2OHmim]Gly) we synthesized not only owns a high CO_2 absorption rate and capacity but also can enhance the antioxidant capacity of the solution, and can be well recycled after multi-cycle regeneration, possesses great potential for making up the deficiencies of alkanolamine, which is beneficial to industrial applications.

(647c) The Effect of Tethering Strategies On the Surface Structure of Amine-Functionalized Ionic Liquids: Inspiration On the CO2 Capture

Xing, H., Zhejiang University
Yan, Y., Zhejiang University
Yang, Q., Key Laboratory of Biomass Chemical Engineering of Ministry of Education, Department of Chemical and Biological Engineering
Bao, Z., Zhejiang University
Su, B., Zhejiang University
Yang, Y., Zhejiang University
Ren, Q., Zhejiang University
Originally presented on: 11/7/2013 9:15:00 - 9:35:00

Amine-functionalized ionic liquids (ILs) are one of the most attractive and promising absorbents for CO_2 capture, because of its high performance and notable advantages over traditional organic amine absorbents. For ILs absorbing CO_2, it is of primary importance to study the structure and orientation of cations and anions at the interface, which presents a great impact on the surface properties and the transportation of CO_2 across the liquid-gas interface. However, despite a variety of cation/anion-tethered amine-functionalized ILs being reported, the choice of the tethering location of the amine group is often somewhat arbitrary and lacks enough consideration in the view of interface chemistry.

Therefore, this work focused on this problem and illustrated the significance of the tethering strategy of the amine functional group on the interfacial structure of ILs for the first time, by investigating the interfacial composition and orientation of cation-/anion-tethered amine-functionalized ILs through molecular dynamic simulation. The results showed that when the amine group is tethered to the cation, strong interaction between the amine group and anions results in a compact interfacial structure without the cation/anion enrichment. However, for amino acid-based ILs with an anion-tethered strategy, the amine group tends to have a higher preference in the outer layer of interface. Therefore, from the viewpoint of interfacial adsorption and transfer, the amine-functionalized ILs with an anion-tethering strategy is considered of great benefit for the CO_2 capture, especially when the amine group is on the terminal side of a relatively long alkyl chain.

(647d) A Carbonate Absorption Process With Crystallization-Enabled High Pressure Stripping for Post-Combustion CO2 Capture: A Kinetic Study and Process Development for Bicarbonate Crystallization

Ye, Q., University of Illinois at Urbana-Champaign
Lu, Y., University of Illinois at Urbana-Champaign
Originally presented on: 11/7/2013 9:35:00 - 9:55:00

A novel carbonate absorption process with high pressure stripping is currently being developed for post-combustion CO_2 capture at the University of Illinois. The process employs an aqueous solution of carbonate, such as K_2CO_3, as a solvent. Bicarbonate in the CO_2-rich absorber effluent is crystallized by cooling the solution from about 70 °C to 30-35 °C in a crystallizer. The resulting bicarbonate slurry is used for CO_2 stripping at a pressure of at least 10 bars. A preliminary techno-economic analysis indicates that the process has an overall energy penalty 30 – 40% lower than that of the bench mark monoethanolamine process, primarily due to lower stripping heat and CO_2 compression work requirements.

The phase transition of bicarbonate during cooling of the CO_2-rich solution is a key feature of the process. A kinetic study of bicarbonate crystallization has been carried out using a laboratory-scale mixed-suspension, mixed-product-removal (MSMPR) reactor. Parametric tests with respect to agitation rate, mean residence time, temperature, and super-saturation level were performed to evaluate their effects on kinetic rates such as crystal nucleation and growth rates. The results were fit to Mydlarz and Jones (1993) model. Empirical regression formulae were then developed to correlate kinetic rates and crystal dimension with process operating conditions. Based on model predictions, mass and energy balance calculations were conducted to predict the performance of a new design configuration employing multiple continuous stirred tank reactors (CSTR) for crystallizing bicarbonate in the proposed CO_2 capture process. This presentation will provide a summary of the kinetic study of $KHCO_3$ crystallization and its implications for the related process design.

(647e) Carbon Capture and Utilization By Three-Step Aqueous Phase Mineralization

Hemmati, A., Sharif University of Technoligy
Bu, J., Institute of Chemical and Engineering Sciences (ICES - A*Star)
Shayegan, J.,
Sharratt, P., A*Star. Institute of Chemical and Engineering Sciences
Originally presented on: 11/7/2013 9:55:00 - 10:15:00

The increase in greenhouse gasses concentration in the atmosphere has contributed to global warming in recent decades. Carbon sequestration is one of the solutions to control the CO_2 concentration in atmosphere. Â A pH-swing mineral carbonation is the process of extracting magnesium or calcium from mineral rock, followed by precipitation of extracted minerals in separate reactors. The process was found to be feasible method for potential industrial application to reduce CO_2 emission by carbon sequestration. This study investigates production of value-added materials for a three-series mineralization reactor in aqueous phase. Magnesium, iron and small portions of some other minerals were extracted from magnesium silicate rock, were precipitated in two steps by increasing the pH from 0.5 up to 10. Product analysis showed that in each step, high purity stable solids such as silica, iron hydroxide II and III, hydrated magnesium carbonate were produced. The CO_2 source was Na_2CO_3 which was the product generated from the scrubbing reaction between a power plant's flue gas and NaOH to serve as a potential carbon capture method. Mass balance for the mineral sequestration process was investigated for future scale up. CO_2 uptake was determined by magnesium carbonate elemental analysis.Â It was calculated that for each ton of sequestrated CO_2 approximately 3.22 ton of serpentine rock (mainly ($Mg_3Si_2O_5(OH)_4$) was required and also 25 m^3 of 1M NaOH, 35 m^3 of 1M Na_2CO_3 and 80 m^3 of 1M HCl were utilized. The resultant value-added products from the process would be 3.1 ton of 99.74% pure nesquehonite ($Mg_2CO_3 \cdot 3H_2O$) and 0.82 ton of iron oxide and hydroxide with 90% purity. Crystalline structure, particle size distribution and morphology of hydrated magnesium carbonate produced, were strongly dependent on the last precipitation reactor conditions including temperature, feeding flow rate, mixing speed and base concentration. Additionally, different composition and crystalline structure of magnesium carbonate produced in the last precipitation step in different temperature ranging 0 to 60 Â°C, shows that the process is capable of producing value-added products including nesquehonite, hydromagnesite and dypingite under a wide range of ambient temperature. Â In future investigations, carbon footprint of the process chain will be studied.Â

(647f) Wood Ash Suspensions in Water As a Solvent in CO2 Absorption Processes

Petho, S. L., Escola PolitÃ©cnica da Universidade de SÃ£o Paulo
Salvagnini, W. M., Polytechnic School of the University of Sao Paulo
Taqueda, M. E. S., Polytechnic School of the University of Sao Paulo
Originally presented on: 11/7/2013 10:15:00 - 10:35:00

Wood ash is a solid waste with organic and inorganic components, obtained from combustion of wood. Its chemical composition may vary according the tree species, soil type, climate and combustion characteristics. The resulting suspension of ash in water has typical properties of high concentrations of carbonates and high pH values. Together with these qualities, the suspension is an interesting solvent in the study of a CO_2 absorption process with low demand of purity and energy.

There are studies about the methodology of determining the calcium carbonate equivalence in ashes because in its composition there other substances like alumino-silicates, silicates and iron oxides that can lead to inaccurate results. So the aim of this study is to develop a simple technique to determine the absorption capacity of unknown ash-aqueous suspensions with carbonates

The experiments were made with wood fuel ashes from a restaurant furnace and validation tests with water and NaOH. The apparatus have a simple characteristics of dismantle all its parts, including the gas tubes and the electrical wirings, forming a called absorption cell, which could be weighted before and after the absorption process.

A simple device was successfully developed that could predict the absorption potential for CO_2 of a solvent with unknown composition in ambient conditions of temperature and pressure. The soluble extracted from *Pterodon emarginatus* Vogel. wood, popularly known in Brazil as Faveiro, had 5.33% of calcium carbonate equivalence. The absorption process with an aqueous solution of this solubles, at aprox. 700 mmHg and 20°C, absorbed 0.77g CO_2 / 100g ashes. Ashes from Eucalyptus has 89.74% of calcium carbonate equivalence and its solubles, 11.01%. The absorption process with the ash-water suspension, at aprox. 700 mmHg and 20°C, absorbed 12.6g CO_2 / 100g ashes.

Analyzing the results of pH in function of time suggested that the absorption of CO_2 occurred because of the total transformation of the carbonates in bicarbonates. This study suggests that wood fuel ashes are a potential solvent for CO_2 absorption.

(647g) CO2 Chemical Transformation to Valuable Commodities: Toward a Utopia for Fossil-Fuel-Based Economies

Taheri Najafabadi, A.,

Originally presented on: 11/7/2013 10:35:00 - 11:00:00

Energy and environment have been always classified as the top human challenges by mid-century. In the contemporary fossil-fuel-based civilizations up to 90% of the energy needs are supplied by the unsustainable resources. Therefore, striking a delicate balance between the versatilities of the fossil fuels and their negative environmental impacts (e.g. global warming and climate change) is crucial. Implementing robust strategies to positively impact global carbon balance ultimately leads to the utopia for the fossil-fuel-based economies; with taking full advantage of carbonaceous fuels while minimizing their undesirable effects. This is where carbon capture and sequestration technologies play a critical role to mitigate CO_2 from the large scale emitters. Current recipes for the CO_2 reduction from large-scale energy consumers (e.g. power stations and cement works) mainly rely on carbon capture and storage (CCS), having three proposed generic solutions: post-combustion capture, pre-combustion capture, and oxy fuel combustion; where the separated carbon dioxide is sent for the physical storage options afterwards. However, researchers believe CCS strategy does not solely resolve the problem permanently, just shifting that from the atmosphere to elsewhere. In fact, long-term ecological impacts of using the earth as a gigantic reservoir for CO_2 are off the beaten track. Adding the potential hazards of CO_2 leakage to the earth surface noticeably exacerbates the prospect of this approach. Considering all the facts, the ultimate solution of realistically facing with CO_2 sequestration concerns is the chemical transformation of CO_2 to valuable commodities such as fuels (through, for instance, Fischer–Tropsch chemistry) or polymers (through successive copolymerization and chain growth). This sustainably reduces carbon emissions, remarkably benefiting from CO_2-derived chemical commodities, so-called carbon capture and conversion (CCC). Nevertheless, CO_2 reduction chemistry is challenging due to the presence of large energy barriers, requiring noticeable catalysis. This work aims to review the most recent advances in the CCC research area selectively (i.e. CO_2 conversion to fuels and CO_2 copolymerization) from chemical engineering viewpoint in terms of both materials and process design. Some of the most promising studies will be presented in detail with the emphasis on subsidizing more research on CO_2 conversion technologies considering the global carbon management urgencies.

(654a) Effects of Inorganic Salts On Absorption of Carbon Dioxide and Oxygen for Alkanolamine Solution Containing Diethyleletriamine and Piperazine in a Rotating Packed Bed

Yu, C. H., National Tsing Hua University
Lin, Y. X., National Tsing Hua University
Tan, C. S., National Tsing Hua University
Originally presented on: 11/7/2013 12:30:00 - 12:55:00

In this work, CO_2 was capture from a gas stream containing 10% CO_2 using an alkanolamine solution containing 15wt% diethylenetriamine (DETA), 15wt% piperazine (PZ) and inorganic salt in a rotating packed bed (RPB). After the study of the addition of potassium chloride (KCl), sodium chloride, lithium chloride, and lithium bromide in the solution on absorption of CO_2 and O_2, KCl was found to be the most effective inorganic salt for the CO_2 capture efficiency, resulting from the longer ionic radius possessed by KCl. Furthemore, the addition of KCl would lower the dissolved oxygen (DO) amount in the solution and yield the same effect as the oxygen scavenger Na_2SO_3. Due to KCl was not consumed by DO, this addition led to no generation of the degradation products created from Na_2SO_3. The effects of gas flow rate, KCl concentration and temperature on CO_2 capture efficiency and DO were examined by a 2^3 factorial design. Gas flow rate was found to be the most significant factor affecting CO_2 capture efficiency followed by temperature and KCl concentration, while temperature was the most significant factor affecting DO followed by gas flow rate and KCl concentration. The optimum operation condition achieving higher CO_2 capture efficiency and lower height of transfer unit in the CO_2 capture process was determined by the response surface methodology. The regeneration energy of 15wt% DETA+15wt% PZ solution containing 1.5 m KCl was found to be lower than that of the solution without KCl since the vapor pressure and heat capacity of the solution were reduced, in a subsequence vaporization heat and sensible heat were reduced as well.

(654b) A Parametric Study of CO2 Absorption Into An Aqueous Potassium Carbonate Solution Promoted With Amines in a Packed Bed Column

Zhang, S.,
Devries, N.,
Lu, Y., University of Illinois at Urbana-Champaign
Originally presented on: 11/7/2013 12:55:00 - 13:20:00

Monoethanolamine (MEA) absorption processes are considered to be state-of-the-art for post-combustion CO_2 capture from coal combustion flue gas; however, they are energy intensive and thus expensive. It is important to develop new advanced CO_2 capture technologies in order to maintain the cost-effectiveness of coal-fired power generation. A novel hot carbonate absorption process with bicarbonate crystallization-enabled high pressure stripping (Hot-CAP) currently under development offers the potential to significantly reduce the energy penalty associated with CO_2 separation from the flue gas and work requirement for CO_2 compression.

In the Hot-CAP, CO_2 is absorbed into a concentrated potassium carbonate (PC) solution (about 40 wt%) at elevated temperatures (60-80°C). Due to the low alkalinity of K_2CO_3, a technical issue to be considered is the low rate of CO_2 absorption into the PC solution compared with that into the MEA solution. Many promoters/catalysts have been studied to enhance the rate of absorption into carbonate solutions in the literatures. However, most of the studies have been limited to low temperatures (≤ 50°C) and low concentrations (<40 wt%) of carbonate solution. For this purpose, different inorganic and organic promoters were screened to enhance the CO_2 absorption rate in the concentrated PC solution at 70°C in our previous work.

In this study, detailed parametric tests, including different dosage of promoter, L/G ratio, and CO_2 loading in solution, were conducted in a bench-scale packed-bed column to evaluate the CO_2 removal performance of the 40 wt% PC solution without and with the additions of three pre-screened amine promoters, diethanolamine (DEA), aminomethyl propanol (AMP), and piperazine (PZ). For the comparison purpose, the performance of CO_2 absorption into a benchmark 5M (30 wt%) MEA was also investigated. Results confirmed that the CO_2 removal efficiency increased with increasing L/G or promoter dosage, and decreased with increasing CO_2 loading. At the same dosage level of promoter, the CO_2 removal efficiency by the PC+PZ was higher than that by the PC+DEA or PC+AMP. Results also revealed that the CO_2 removal efficiency by the 40 wt% PC solution promoted with 1M DEA or 0.5M PZ, either for the CO_2 lean or rich solution, was higher than that of the 5M MEA counterpart solution under their typical operating conditions.

This presentation will provide a summary of results from the parametric study of CO_2 absorption into the concentrated PC solution promoted with three pre-screened amine promoters.

(654c) Prediction of Reaction Rate Constant of Carbon Dioxide Absorption Into Aqueous Amine Solution

Goto, K., Research Institute of Innovative Technology for the Earth
Yamada, H.,
Higashii, T., Research Institute of Innovative Technology for the Earth (RITE)
Originally presented on: 11/7/2013 13:20:00 - 13:45:00

Abstract:

Carbon dioxide (CO_2) capture from a gas stream is a state-of-the-art technology and has been still developed to implement CO_2 capture and storage (CCS), which is a promising countermeasure of global warming. This study proposes a prediction scheme of a reaction rate constant for CO_2-amine-H_2O reaction system in order to provide it to process evaluation of CO_2 capture with an aqueous amine solution. Firstly, gas scrubbing tests for twelve aqueous amine solutions including primary, secondary, tertiary and hindered amines were conducted, then absorption behaviors obtained from the experiments were discussed with consideration of basicity (pKa) and reaction kinetics.

CO_2 absorption rates of aqueous amine solutions were measured through the screening test using a gas scrubbing bottle of 250 ml volume. A gas mixture of CO_2 and N_2 was supplied to the bottle with 50 ml of an aqueous amine solution at a flow rate of 700 ml/min for 60 minutes at the atmospheric pressure. The flow rate and CO_2 concentration of the feed gas were constant during the test and the CO_2 concentration of outlet gas from the bottle was analyzed with IR Analyzer. The CO_2 loading and absorption rate with time were calculated from the outlet gas CO_2 concentration. The absorption rate (N_{CO2}) shown in the test represented a relative CO_2-absorption rate among solvents. Test condition of CO_2 concentration was 20%, which was a model gas concentration of blast furnace gas in the steel industry. Aqueous amine solution was adjusted at amine concentrations of $2mol_{-amine}/L$.

As a result, the absorption rate (N_{CO2}) had a strong correlation with reaction rate constant of CO_2-amine second order reaction rate constant, which was cited from previous published literatures. The following linear equation was given for aqueous amine solutions investigated in this study (primary, secondary and hindered amines).

$$\log(k_2)=a\,N_{CO2}+b$$

In the case of the experimental condition (solvent: 2mol_{-amine}/L, gas: 20%$_{-CO2}$) of this study, constants a / b were 0.0107 / -0.3726 for hindered amines. AMP (2-amino-2methyl-1-propanol), AMPD (2-amino-2metyl-1,3-propanediol), AMHD (2-amino-2-hydroxymethyl-1,3-propanediol) and 2PE (2-piperidineethanol) were considered as the hindered amines here.

Furthermore, it was shown that correlation equation between reaction rate constant and pKa for hindered amines could be expressed with the following equation,

$$\ln(k2)=c1\ pKa + c2 - c3 / T.$$

In brief, the scheme for prediction of reaction rate constant from the simple gas scrubbing test could make it easy to evaluate CO_2 capture with an aqueous amine solution and to accelerate development of high performance CO_2 capture solvent.

Acknowledgments

This work was financially supported by the Ministry of Economy, Trade and Industry (METI), Japan.

(654d) Modeling the Absorption of CO2 Into Aqueous Blends of Deea and Mapa

Garcia Moretz-Sohn Monteiro, J., Norwegian University of Science and Technology
Hammad, M., NTNU
Hussain, S., NTNU
Knuutila, H., Norwegian University of Science and Technology
Svendsen, H. F., Norwegian University of Science and Technology

Originally presented on: 11/7/2013 13:45:00 - 14:10:00

Introduction

The reaction rate of CO_2 absorption by amine solutions is a key parameter for designing an absorption column. The faster the mass transfer, the lower the contact area needed and the closer to equilibrium one may get. Tertiary amines promote the reaction of hydration of CO_2 leading to bicarbonate formation (Donaldson and Nguyen, 1980), a slow reaction, while primary and secondary amines form carbamates, a relatively fast reaction. Blending different types of amines has been shown to be a promising path in solvent development for CO_2 capture by absorption. If there is synergy between the different properties of the amines, this leads to an enhanced process performance.

Aqueous blends of N,N-diethylethanolamine (DEEA), a tertiary alkanolamine, and 3-(methylamino)propylamine (MAPA), a diamine with one primary and one secondary group, are especially interesting for the absorption process because they form two liquid phases upon CO_2 loading. The upper phase is lean in CO_2 and does not require regeneration, while the lower phase is CO_2-rich and is sent to the stripper column. There, DEEA is mainly regenerated (Liebenthal et al., 2012). The process can therefore combine lower energy requirements in the regeneration side and high reaction rate in the absorption side of the process.

The combined reaction rate of CO_2 absorption between DEEA and MAPA is presented in this work.

Experiment

The reactive absorption of CO_2 into aqueous solutions of DEEA and MAPA was conducted in a string of discs contactor (SDC), previously described by Knuutila *et al.* (2009) among others. A gas stream, containing N_2 and CO_2, and the aqueous amine solution were contacted in countercurrent mode.

The inlet flow rates of N_2 and CO_2 were set using mass flow controllers, while the CO_2 composition in the outlet gas was determined by a CO_2 analyzer. There were indicators for inlet and outlet temperatures of the liquid and gas streams. As a surplus, the inlet liquid flow rate was controlled. The experiment was set so that the reaction follows the pseudo-first order regime.

The reaction was performed within the temperature range 25°C to 60°C for unloaded aqueous blends of DEEA and MAPA with varying composition. In total, 10 blends were tested. Some physical properties of the systems, namely density, viscosity, vapor pressure, Henry's law constant, were measured to enable the modeling.

Modeling

The rate of CO_2 absorption and the overall mass transfer coefficient (K_{OV}) could be directly calculated from the experimental data. The knowledge of the physical properties of the system (density, viscosity, vapor pressure, Henry's law constant) enabled an evaluation of the liquid and gas-film mass transfer coefficients (k_L and k_G), as well as initial reaction rate constant k_{obs}.

The reactions involved in the absorption of CO_2 by aqueous blends of DEEA and MAPA are described as:

$CO_2+OH^- \rightarrow HCO_3^-$ r_1

$DEEA + H_2O + CO_2 \rightarrow DEEAH^+ + HCO_3^-$ r_2

$MAPA + H_2O + CO_2 \rightarrow MAPACOO^- + H3O^+$ r_3

Following the approach of Knuutila *et al.* (2009), the rate equations are written in terms of activities, so that it is consistent with rigorous VLE models in which the equilibrium constants are given in terms of activities. Taking the pseudo-first order assumption into consideration, the forward CO_2 reaction rate is given by:

$-r_{CO2} = r_1+r_2+r_3 = k_{obs}a_{CO2}$ eq.1

The initial rate constant can then be modeled in terms of the rate constants of the 3 reactions defined above (k_{OH} and k_{DEEA} and k_{MAPA}):

$kobs = k_{OH}*a_{OH-} + k_{DEEA}*a_{DEEA}+k_{MAPA}*a_{MAPA}$ eq.2

where reactions 1 and 2 are modeled as second-order reactions. If reaction 3 is modeled based on the direct (termolecular) mechanism (da Silva and Svendsen, 2004), considering MAPA, water and DEEA are considered as dominating bases, k_{MAPA} is described as:

$k_{MAPA}*a_{MAPA} = k'_{DEEA}*a_{DEEA} + k'_{MAPA}*a_{MAPA} + k_{H2O}*a_{H2O}$ eq.3

The CO_2 activity is given by the experimental Henry's Law constant, while the activities of the other species are calculated using the NRTL model parameters presented by Hartono et al. (2013)

Acknowledgements

Financial support from the EC 7th Framework Programme through the iCap project, Grant Agreement No : iCap-241391, is gratefully acknowledged.

References

da Silva, E.F.; Svendsen, H. F. *Ab Initio Study of the Reaction of Carbamate Formation from CO2 and Alkanolamines*. Ind. Eng. Chem. Res, 2004, 43, 3413-3418

Donaldson, T.; Nguyen, Y. N. *Carbon Dioxide Reaction Kinetics and Transport in Aqueous Amine Membranes*. Ind. Eng. Chem. Fundam. 1980, 19, 260-266.

Hartono, A., Saleem, F., Arshad, M.W., Usman, M., Svendsen, H.F., 2013. *Binary and ternary VLE of the 2-(diethylamino)-ethanol (DEEA)/ 3-(Methylamino)-propylamine (MAPA)/ Water system*. Submitted to the International Journal of Greenhouse Gas Control.

Liebenthal, U.; Pinto, D.D.D.; Monteiro, J. G. M.-S.; Svendsen, H. F.; Kather, A. *Overall process analysis and optimization for CO2 capture from coal fired power plants based on phase change solvents forming two liquid phases*. 11th International Conference on Greenhouse Gas Control Technologies, 18 - 22 November 2012, Kyoto, Japan

Knuutila, H.; Svendsen, H. F.; Juliussen, O. *Kinetics of carbonate based CO_2 capture systems*. Energy Procedia. 2009, 1, 1011-1018.

(654e) Pilot Test and Simulation of An Advanced Amine Process for CO2 Capture
Chen, X., The University of Texas at Austin
Originally presented on: 11/7/2013 14:10:00 - 14:35:00

Pilot Test and Simulation of an Advanced Amine Process for CO_2 Capture

Xi Chen, Barath Baburao, Frederic Vitse*

Alstom Power, 1409 Centerpoint Blvd, Knoxville, TN 37932

An Advanced Amine Process (AAP) was extensively tested in the Alstom Power pilot plant facility located at Växjö, Sweden. The performance of a proprietary solvent X was evaluated under varied flow schemes.

The test campaign comprised several series of tests, including energy consumption at varied ratio of liquid to gas flow rate in absorber with 90% CO_2 removal, effect of absorber packing height, and energy performance under advanced flows schemes as compared to conventional ones. With advanced flow schemes, the energy demand for CO_2 separation excluding CO_2 compression work was reduced by 13% - 21% compared to the conventional flow scheme, at 90% CO_2 capture.

The predictive capability of Procede Process Simulator was examined against the pilot plant data for absorber and regenerator. The correlations for mass transfer and interfacial area developed by The University of Texas at Austin were implemented for the rate-based modeling of Mellapak 2X absorber packing. The comparison between the simulation results and the pilot plant data shows that the model is capable of satisfactorily predicting the performance of AAP under different process conditions and configurations. Comparison of the simulation to the experimental results in terms of deviation for key parameters is summarized in the following Table.

Table 1. Deviation between simulation and experimental results for AAP performance.

Predicted Variable	Average Absolute Deviation (AAD[a])	BIAS[b]
Absorber CO2 removal percentage (wrt. Liquid)	4.7%	-0.8%
Absorber outlet amine solvent flow rate	1.1%	-0.5%

Absorber rich loading	2.7%	-0.4%
Stripper lean Loading	2.8%	-0.2%
Stripper outlet gas flow rate	14.8%	-6.5%
Stripper gas temperature at the outlet of the column (K)	3.4[c]	-1.3 [c]
Stripper outlet amine solvent temperature (K)	0.4[c]	0.3 [c]

a: $AAD = \frac{1}{n}\sum_{i=1}^{n}\left|\frac{y_{i,calc} - y_{i,exp}}{y_{i,exp}}\right| \times 100\%$

b: $BIAS = \frac{1}{n}\sum_{i=1}^{n}\frac{y_{i,calc} - y_{i,exp}}{y_{i,exp}} \times 100\%$

$n =$ number of data points

c: For temperature, the arithmetical mean value of deviation is used.

* Corresponding author. Tel.: +1-865-560-1732; fax: +1-865-560-1642.

E-mail address: frederic.vitse@power.alstom.com.

(654f) Techno-Economic Analysis for CO2 Capture and Storage Processes With Concentrated PZ and Mdea/PZ Blends

Frailie, P. T. II, The University of Texas at Austin
Sachde, D. J., The University of Texas at Austin
Madan, T., The University of Texas at Austin
Rochelle, G. T., The University of Texas at Austin
Originally presented on: 11/7/2013 14:35:00 - 15:00:00

This study proposes a method for quantifying the relationships between capital and operating costs for retrofit alkanolamine scrubbing systems for the removal of CO_2 from the flue gas of coal-fired power plants. This particular application was chosen because of its potential to significantly reduce CO_2 emissions with relatively little energy penalty. Coal-fired power plants generate forty-five percent of the electricity in the United States and about seventy-seven percent of the CO_2 from the power sector, and the relatively high concentration of CO_2 reduces the energy required to capture 90 % of the greenhouse gas emissions. Concentrated piperazine (PZ) and methyldiethanolamine (MDEA)/PZ blends were chosen because they exhibit favorable rates and capacities over operationally significant conditions. Using an Aspen Plus® thermodynamic, hydraulic, and kinetic model developed at The University of Texas at Austin, a base-case absorption/stripping/compression process was designed, tested, and optimized to minimize the cost of CO_2capture and storage.

The Aspen Plus® thermodynamic model was developed using a sequential regression methodology to regress CO_2 solubility, heat capacity (loaded and unloaded), CO_2 activity coefficient, amine volatility (loaded and unloaded), amine pKa, and heat of absorption over operationally significant temperature, amine concentration, and loading ranges for MDEA, PZ, and MDEA/PZ systems using the electrolyte-NRTL method. Hydraulic data was directly incorporated into Aspen Plus® using FORTRAN subroutines that calculate viscosity, density, and binary diffusivities for amine-derived species as a function of process parameters such as temperature, amine concentration, and CO_2 loading. CO_2 absorption rate data was incorporated by adjusting rate constants and binary diffusivities in an Aspen Plus® wetted wall column (WWC) simulation and a Microsoft® Excel regression to fit experimental data at various temperatures and loadings. The final kinetics model predicts the CO_2 absorption rate for 5 m PZ, 8 m PZ, 7 m MDEA/2 m PZ, and 5 m MDEA/5 m PZ from 10-100 °C over operationally significant loading ranges with an average error of 4.9 %.

The techno-economic analysis focused on four major cost centers: (1) absorber, (2) cross-exchangers, (3) reboiler, and (4) compressor. These four unit operations can constitute as much as 85 % of the capital cost, and, thus, present the best opportunity for reducing the cost of CO_2 capture. Reducing the cost of one unit operation typically results in increasing the cost of another. For example, increasing the liquid flow rate will decrease the required packing volume (lower CAPEX), but it will increase the size of the cross-exchangers (higher CAPEX) and the reboiler duty (higher OPEX). It is possible that the savings in the absorber size will exceed the losses in cross-exchanger size and reboiler duty, but this cannot be quantified without a rigorous techno-economic analysis. Expressions have been developed for the four aforementioned unit operations that calculate CAPEX and OPEX in the units of US dollars per metric ton of CO_2 using Aspen Plus® predictions for stream properties. These expressions are based on price quotes obtained from manufacturers, previous techno-economic analyses for alkanolamine CO_2 capture processes, and Aspen Plus® ICARUS. CAPEX predictions were converted to the units of US dollars per metric ton of CO_2 using a factor that accounts for both the conversion of purchased equipment cost to total capital requirement and the annualizing of the initial investment over the lifetime of the project. This factor was derived using pricing guidelines from non-specific estimation methods such as chemical engineering design texts, as well as CO_2-capture specific guidelines from the Department of Energy and the IEAGHG.

The base-case absorber design will employ either a pump-around or an in-and-out intercooler. Pump-around intercooling effectively splits the column into three sections: (1) a top section into which lean solvent enters and scrubbed gas leaves, (2) a middle section containing 2-5 times more solvent than the top section, and (3) a bottom section containing the same amount of liquid as the top section from which the rich solution exits and the flue gas enters. In-and-out intercooling simply removes all of the solvent from the top section, cools it to 40 °C, and feeds it to the top of the bottom section. There are tradeoffs associated with lean loading, feed liquid flowrate, and intercooling configuration. Decreasing lean loading increases capacity and decreases absorber height but increases reboiler duty. Increasing the feed liquid flowrate from 1.1 to 1.2 times the minimum liquid flowrate decreases the capacity of the solvent by 8 to 11 % but decreases the amount of packing needed to remove 90 % of the CO_2 by 25 to 30 %. The base-case stripper contains both a cold rich bypass (CRB) and warm rich bypass (WRB) stream. A fraction of the cold rich solvent exiting the bottom of the stripper is heated by the product gas before being fed into the top of the stripper. The remaining rich solvent is heated by a warm solution from the bottom of the stripper in a cross exchanger. Another portion of the warm rich solvent is bypassed and fed directly into the top of the stripper. The remaining rich solution is heated by a steam heater and flashed into the bottom of the column.

After developing the pricing method and designing a base-case configuration, operating conditions such as lean loading, solvent flow rate, and intercooling configuration were varied to minimize the cost of CO_2 capture. For each condition the percent bypass to the CRB and WRB was optimized to minimize the equivalent work, absorber section heights were adjusted to minimize packing area, heat-exchanger sizes were optimized for heat transfer area and pressure drop, and column diameters were adjusted to achieve a maximum of 70 % flood. This analysis was performed for 8 m PZ, 7 m MDEA/2 m PZ, and 5 m MDEA/5 m PZ.

(702a) Low-Pressure Sweep Membrane Modules for Carbon Dioxide Capture

Kniep, J., MTR
Hao, P., MTR
Chan, K., MTR
Fulton, D.,
Freeman, B., Electric Power Research Institute
Merkel, T., MTR
Baker, R., MTR
Originally presented on: 11/7/2013 12:30:00 - 12:55:00

To mitigate the harmful effects of global climate change, CO_2 in power plant flue gas must be captured and either sequestered or utilized in processes such as enhanced oil recovery. Membrane technology is an attractive CO_2 capture option because of advantages such as energy-efficient passive operation, tolerance to acid gases and oxygen, no hazardous chemical emissions or handling issues, minimal water requirements, and no steam use requiring modifications to the existing boiler.

Working with DOE, MTR has previously demonstrated the feasibility of a two-step membrane process design to recover CO_2 from power plant flue gas. The second step of this process is a selective sweep membrane module that strips CO_2 from the flue gas and recycles it back to the boiler with the incoming combustion air. MTR is developing a membrane module specifically designed for high gas flow, low pressure, sweep operation. These membrane modules contain up to 500 m^2 of membrane area, which is 10 to 20 times the membrane area of modules currently being used for flue gas CO_2 capture. These large area modules will reduce the manifolding complexity, footprint, and cost of the very large membrane plants required for flue gas treatment. By lowering the pressure drop through the modules, less compression/blower energy is needed to move gases through the module, which reduces the parasitic energy loss of the capture process. In this presentation, results of the development work to date and future plans, including slip stream field testing at the National Carbon Capture Center (NCCC) operated by Southern Company in Birmingham, AL, will be discussed.

The development of large, efficient, low cost, sweep membrane modules is an enabling technology for membrane CO_2 capture and sequestration systems. MTR is investigating several alternative process designs to remove CO_2 from coal power plant flue gas, and while all of the designs use a membrane contactor to recycle CO_2 to the power plant boiler, the second CO_2 separation unit need not be a membrane unit. The selection of different hybrid separation technologies, hybrid process designs, and process economics will be presented.

(702b) Major Interest of Hollow Fiber Membrane Contactors for An Ammonia-Based CO2 Capture Process: Detailed Analysis of An Innovative Process

Makhloufi, C.,

Roizard, D., Nancy Université

Favre, E., Laboratoire Réaction et Génie des Procédés LRGP- CNRS, Université de Lorraine

Originally presented on: 11/7/2013 12:55:00 - 13:20:00

Up to now, the reference technology for CO_2 capture in postcombustion makes use of monoethanolamine (MEA) as chemical solvent in a packed column. Despite high capture efficiencies, MEA induces too high energy penalties. Consequently, an abundant literature is discussing the use of aqueous ammonia for flue gas scrubbing. Indeed ammonia is cheaper than MEA, shows a higher absorption capacity and, most of all, its energy regeneration demand is much lower. Nevertheless, the high ammonia volatility is a considerable drawback which causes a high NH_3 slip within the flue gas. Specific operatic conditions are required to limit NH_3 losses and a post absorber scrubbing step is therefore needed.

Amongst growing technologies, membrane contactors appear as a promising technology able to allow significant process intensification when compared to conventional gas/liquid contactors. So far, no study dealing with the use of membrane contactors for the CO_2 capture in aqueous ammonia has been disclosed. For the first time, a detailed analysis of an ammonia based CO_2 capture process using hollow fiber membrane contactors will be given during this presentation. Particularly, membrane contactors ability to both improve CO_2 mass transfer performances and mitigate NH_3 volatilization compare to conventional gas/liquid contactors will be discussed.

First of all, it will be shown that CO_2 absorption in ammonia can not be performed using conventional microporous hollow fiber membrane contactors and composite membranes made of a thin polymeric layer facing the liquid phase must be used. In that sense, composite hollow fiber membrane contactors have been specifically designed, prepared and tested to favor CO_2 absorption and, in the same time, to limit the ammonia slip in the flue gas.

This involved precisely selecting the dense layer and it will be shown how an inadequate material selection could lead to its significant degradation. The chosen dense skin must be thin, highly permeable to CO_2 and selective for CO_2 over NH_3. This should allow reaching high CO_2 mass transfer coefficient while reducing drastically the ammonia slip in the flue gas. This was undoubtedly one of the main challenges. Indeed, it has to be recalled that a faster permeation of CO_2 compared to NH_3 in dense polymers was totally unexpected and to our knowledge unexplored. For the first time, 4 chemically resistant polymers showing a reverse selectivity CO_2/NH_3 were identified. These meaningful results obtained thanks to time lag experiments will be presented in this presentation.

In a second step, CO_2 absorption experiments in aqueous ammonia using lab scale composite hollow fiber membrane contactors (CHFMC) have been successfully carried out. The effect of the main operating conditions namely ammonia concentration, temperature, gas and liquid flow rates, nature of the dense layer, on the CO_2 absorption performances and the ammonia loss has been studied. The overall volumetric CO_2 absorption capacity in CHFMC has been compared to the one obtained in packed columns. It has been concluded that membrane contactors strongly enhance the CO_2 absorption performances in comparison to conventional contactor. Moreover, a sharp decrease of the ammonia slip has been observed. The above results will be detailed in the presentation and the performances and advantages of this new CO_2 absorption process will be discussed further.

(702c) Membrane Assisted Liquid Absorbent Regeneration (MALAR):Innovative Use of Membranes in Liquid Absorbent Based CO2 Capture

Feron, P.,
Zhao, S.,
Originally presented on: 11/7/2013 13:20:00 - 13:45:00

CO_2 capture and geological storage (CCS) is a necessary option to limit CO_2 emissions from existing power stations. In the CCS chain the capture and compression of CO_2 to pipeline or injection pressure represents around 70% of the costs and all of the energy requirements of the process. Typical output reductions for 90% capture of CO_2 at a power station are around 30% and these are very large indeed. The leading capture technology is based on a liquid absorbent which chemically reacts with CO_2. The liquid absorbent is regenerated thermally and it will be re-used during many cycles. The CO_2produced from this process is also of high purity and can easily be stored in the underground after compression.

The energy required for the liquid absorbent regeneration is extracted as steam from the power station. This steam is then not available for power generation. Hence the output of the power station is reduced, which is undesirable. It is therefore necessary to develop absorbent regeneration processes which have much lower energy requirement. The starting point is to analyse the thermal energy requirement in more detail. It has the following contributions:

- The energy required to break the bond between CO_2 and the active component in the solvent as determined by the chemical formulation of the liquid absorbent.
- The heat required for bringing the solvent up to the regeneration temperature, which is a function of the CO_2 loading of the liquid absorbent.
- The evaporation enthalpy for the stripping steam which leaves the stripper together with the CO_2, which is determined by the temperature of the steam-saturated CO_2-product leaving the stripping column.

The overall energy penalty also incorporates the electricity use of the capture and compression process. This penalty is however to a large extent determined by the heat required to release the CO_2 from the liquid absorbent, typically around 4 MJ/kg CO_2. Dependent on the fuel type, this represents between 20 and 40% of the energy contained in the incoming fuel.

The ongoing development of new liquid absorbents will very likely reduce the first two contributions, i.e.

new amines with lower binding energy and higher CO_2-loadings have been identified. However the energy required for the production of stripping steam is still needed. In fact all the energy needed for the regeneration is brought in via the stripping steam and is transferred to the liquid absorbent via condensation. Hence efficient control of the steam balance is important for the energy performance of the capture process.

The MALAR process builds on this foundation, aiming to reduce the energy requirement to less than 2 MJ/kg CO_2 using a **M**embrane **A**ssisted **L**iquid **A**bsorbent **R**egeneration process. This process concept lends itself for application with any type of liquid absorbent.

The proposed innovation aims to efficiently integrate heat recovery in the liquid absorbent regeneration process. This is achieved by the use of membranes for transfer of heat using condensation and evaporation processes. The process concept is an entirely new one utilising membranes as gas-liquid contactors in such a way that the (common) occurrence of liquid leakage through the membrane is not an issue.

In the MALAR process energy is recovered via transfer of steam through a membrane. In the membrane condenser latent heat is recovered from the wet CO_2 product into a portion of the cold CO_2-rich liquid absorbent; in the membrane evaporator latent heat is recovered from the hot CO_2-lean absorbent into the stripping column. The use of membranes for direct heat transfer via evaporation or condensation enables quite small approach temperatures which would be unachievable and/or uneconomical with conventional heat exchangers. The compactness of the membrane equipment further adds to the achievement of small temperature differences and thereby a very energy efficient process. Two porous membrane configurations using either hyrodophobic or hydrophilic are proposed for the exchange of latent heat.

This application of membranes is a so-called membrane contactor application, i.e. the membrane is used to contact a gas flow with a liquid flow. Such operations often suffer from minor amounts of liquid permeating through the membrane to the gas phase. This liquid is subsequently carried away by the gas flow, which is generally not desired, and particularly if the gas enters the atmosphere. It will furthermore hamper the efficient transfer of components between the two phases. In the MALAR application envisaged here, liquid leakage is not critical as the heat transfer function is not affected and the liquid will be contained in the stripper column and mixed with the liquid absorbent already present.

A 50% reduction in thermal energy required for the liquid absorbent regeneration can be achieved by using the MALAR concept in combination with improved liquid absorbents, resulting in a thermal energy requirement below 2 MJ/kg CO_2. The improvements are coming from:

- A lower gas temperature exiting the stripper column reducing the evaporation loss
- A closer temperature approach through the use of the membrane heat exchangers
- The liquid absorbent having a double CO_2-loading and lower reaction enthalpy, which a feature of new liquid absorbents developed over the last 2 decades.

In summary: the benefits of the MALAR are:

- Energy efficient capture process through novel heat recovery method
- Smaller size equipment through use of high specific surface area membrane equipment
- Applicable to a wide range of liquid absorbents
- Different membrane types (hydrophobic and hydrophilic) are feasible
- Use of a membrane contactor in a non-critical application with respect to liquid leakage

(702d) Pressure Swing Membrane Absorption for Pre-Combustion CO2 Capture From Shifted Syngas

Sirkar, K. K., New Jersey Institute of Technology
Chau, J., New Jersey Institute of Technology
Jie, X., New Jersey Institute of Technology
Obuskovic, G., New Jersey Institute of Technology
Originally presented on: 11/7/2013 13:45:00 - 14:10:00

An advanced pressure swing membrane absorption (PSMAB) technology utilizing a hollow fiber membrane contactor device and a cyclic absorption-desorption process is being developed to produce purified hydrogen at a high pressure for IGCC-CCS plant's combustion turbine from low temperature (L-T) post-shift reactor synthesis gas and simultaneously obtain a purified CO_2 stream suitable for sequestration. The low temperature water gas shift reactor product stream is likely to be available at ~20 atm and around 150-200^0C. The experimental studies are employing He as a surrogate for H_2. Studies have been carried out with a 40%CO_2-60% He feed gas mixture coming in at temperatures up to 100^0C and pressures varying from 100 to 250 psig (689-1723 kPag). The nonvolatile absorbent liquid consists of an ionic liquid 1-butyl-3-methylimidazolium dicyanamide ([bmim][DCA]) (among others) containing nonvolatile amines. Two different cyclic processes have been studied; most of the performance will focus on a 5-valve cycle. Process performance studies have focused on the duration of various parts of the cycle, the effects of pressure and temperature variation, and the effect of hollow fiber membrane module design on the quality of the two product streams namely, the He-rich stream and the CO_2-rich stream. A numerical model has been developed to predict the process performance of the simpler 3-valve cycle for pure ionic liquid as the absorbent. Solubility and diffusion coefficient data for the gases are needed in this model; these were determined for the pure ionic liquid absorbent using a pressure-decay-dual transducer apparatus at temperatures of 323, 353, 363, and 373 K and at pressures up to 1.38 MPa (~200 psig).

(702e) Application of Composite Polymeric Membranes for Efficient Regeneration of Physical Solvents

Pesaran, A.,
Tande, B., University of North Dakota
Originally presented on: 11/7/2013 14:10:00 - 14:35:00

The most dominant CO_2 capture technology used for pre-combustion capture involves the application of physical solvents. Unlike chemical solvents, physical solvents do not react with the solute and they physically dissolve the acid gases, which are then stripped without the need to be heated by means of pressure swing techniques or combination of heat and pressure letdown. However, despite low energy required to regenerate the physical solvents and their high capacity to capture and separate acid gases from the syngas produced in a gasification plant, physical solvents have some disadvantages including CO_2 pressure loss and the energy required to pump the solvent to the high pressure absorber.

The primary objective of this work is to evaluate the use of composite polymeric membranes, for the recovery of CO_2 from CO_2-rich solvent streams. To achieve this purpose, an experimental bench-scale setup was built to investigate and quantify CO_2 removal capacity from the rich solvent across different types of membranes. This setup primarily consists of a high pressure absorber where solvent is loaded and saturated with CO_2 and the membrane module where CO_2 diffuses through the membrane. In order to increase the driving force across the membrane, sweep gas is used (Nitrogen). To evaluate the absorption process, absorber temperature and pressure is measured and recorded continuously using a NI data logger and the LabView software. The absorber is equipped with a home-made cooling water coil and a relief valve to adjust the absorber temperature and pressure respectively. Two 400 W inline pencil heaters are used and controlled with a temperature controller to adjust the solvent temperature downstream of the membrane module. The solvent line temperature and pressure are also recorded continuously. CO_2 concentration in the sweep gas stream is measured using the Non-Dispersive Infrared CO_2 analyzer and Agilent 7850A GC- FID depending on the concentration of CO_2 in the sweep gas. Dimethyl ether of polyethylene glycol (Selexol) is used as the solvent since it is reputed to be one of the major physical solvents for CO_2 removal. To evaluate the effectiveness of different types of membranes, CO_2 permeation rate and membrane selectivity were measured for different membranes. To better understand the effect of system pressure and solvent flow rate on different experiment responses such as CO_2 flux, selectivity and % recovery, a two-factor two -level full factorial design with two replicates an three center points were performed with the membrane that showed the most promising results in the screening study.

The results of the screening study indicated that PDMS based membranes (PERVATECH and PERVAP

4060) shows higher CO_2 permeability compared to PVAOH based membranes (PERVAP 1211 and PERVAP 1201). The best membrane for further analysis and experiments to find the optimum operational conditions was chosen as PEVAP 4060 from SULZER due to its high CO_2 flux and selectivity compared to other membranes.

A statistical analysis was performed to identify the significant factors for each individual response such as permeation rate, leak rate and selectivity. For CO_2 flux, pressure appeared to be strongly significant. However, solvent flow rate had no significant effect on the rate of CO_2 permeation. The immediate conclusion from this observation is that the main mass transfer is controlled by the membrane. With respect to the solvent leak, the analysis of Pareto charts suggested pressure to be significant and solvent flow rate to be insignificant. As the system pressure increases, the liquid on the upper chamber of the membrane module forces itself more into the membrane and hence, solvent leak increases. However, increasing the solvent flow rate leads to higher liquid velocity on top of the membrane and shows no significant changes on the rate of solvent leak. Neither system pressure nor solvent flow rate found to be significant considering the selectivity as the experiment's response. Finally, regarding the percent recovery, both the system pressure and solvent flow rate appeared to be significant. This is mainly because at elevated pressures, mole fraction of CO_2 in solvent increases and thus a higher driving force for CO_2 permeation exists. As a result of this, percent of recovery increases by pressure as confirmed by the Main Effects plots. On the other hand, by increasing the solvent flow rate, more amount of CO_2 enters the upper membrane chamber but the mass transfer resistant through the membrane prevents more CO_2 from being transported. Thus, there is the condition of introducing more CO_2 to the upper chamber but still a slow mass transport process being controlled by the membrane which eventually decreases the separation ability of the membrane (percent recovery).

In order to examine the chemical stability and structural integrity of the membranes after being exposed to the high pressure solvent, a series of post-experiment characteristic tests such as FTIR and DSC were performed. The results of these studies revealed no major changes.

(702f) Efficient Capture of CO2 From Pulverized Coal Fired Plants

Dada, E. A., ChemProcess Technologies (CPT), LLC
Olanrewaju, K., University of Iowa
Drayton, R. A., THS, Inc
Originally presented on: 11/7/2013 14:35:00 - 15:00:00

As demand for fossil fuel based energy increases, the need to sustainably manage and mitigate carbon dioxide (CO_2) emissions, the largest volume contributor to global warming, is unequivocally necessary. Various studies discussing different methods of capturing/sequestrating $CO_{2,}$ either by separation from a stream of effluent flue gases after a combustion process or through fixation, have been thoroughly reviewed. These methods range from chemical and physical absorption and adsorption, to forestation, Ocean fertilization, photosynthesis process, mineral carbonation and other *micro*-scale sequestration schemes. As this study will focus more on the technical options of separating CO_2 from pulverized coal-fired plant (a *post* combustion process), the current method of chemically absorbing CO_2 from stream of flue gases is fraught with challenges such as: the energy intensity of the process, limited loading capacity, solvent degradation and the corrosive effect of the solvent upon the process units. While physical absorption and adsorption address some of the limitations within chemical sorption process, the shortcomings still persist in their inability to handle large volume of flue gases with low CO_2 partial pressures. This work discusses two novel solutions that address the issues concerning the current technical approaches of sorbing carbon from effluent flue gases at coal-fired plants. It is proposed to design an energy efficient process that is cost effective and novel solvent formulation that an enhance CO_2 loading capacity due to their higher chemical stability and lower corrosive characteristics.

(707b) CO2 Capture and Conversion Based On ILs

Zhang, S., Institute of Process Engineering, Chinese Academy of Sciences
Originally presented on: 11/7/2013 12:55:00 - 13:20:00

CO_2 Capture and Conversion Based on ILs

Suojiang Zhang*, Jian Sun

Institute of Process Engineering, Chinese Academy of Sciences,100190, Beijing, PR China

**Corresponding author: sjzhang@home.ipe.ac.cn*

An ionic liquid based CO_2 capture technology has been developed. To enhance the absorption efficiency, a series of ILs were synthesized and screened including amino functionalized IL.[1] The performance of IL mixture for CO_2 capture was also carried out.[2]The system integration result indicated that the new process resulted in about 27% reducing of energy consumption comparing to the traditional amine process.

As a way out for CO_2 capture, the synthesis of cyclic carbonates via cycloaddition of CO_2 to epoxides is studied, where the synthesized ILs are found to be very efficient catalysts for the cycloaddition reaction. Our previous work indicated that the hydroxyl group in solvent, IL or carrier all could play a promotion role on the reaction. [3] Based on the fundamental understanding, hydroxyl or carboxyl-functionalized ILs were developed and found to be much more active.[4] Furthermore, these ILs are chemically immobilized on solid supports, including silica, resin, zeolites, and chitosan, which show high stability and efficiency.[5]

Refferences

[1] a) X. P. Zhang, X. C. Zhang, H. F. Dong, Z. J. Zhao, S. J. Zhang, Y. Huang. ***Energy Environ. Sci.***, 2012, 5: 6668-6681. b) J. M. Zhang, J. Sun, X. C. Zhang, Y. S. Zhao, S. J. Zhang. ***Greenhouse Gas Sci Technol.***, 2011, 2:142-159.

[2] a) Y. S. Zhao, X. P. Zhang, Y. P. Zhen, H. F. Dong, G. Y. Zhao, S. J. Zeng, X.Tian, S. J. Zhang. ***Int. J. Greenh. Gas Con.***, 2011, 5:367-373; b) Y. S. Zhao, X. P. Zhang, H. F. Dong, Y. P. Zhen, G. H. Li, S. J. Zeng, S. J. Zhang. ***Fluid Phase Equilib.***, 2011,302: 60-64.

[3] a) J.Q. Wang, J. Sun, W.G Cheng, S.J. Zhang. ***Phys. Chem. Chem. Phys.***, 2012, 14, 11021-11026; b) J.

Sun, J. Y. Ren, S. J. Zhang, W. G. Cheng. ***Tetrahedron Lett***., 2009, 50(4): 423-426.

[4] a) X. Chen, J. Sun, J. Q. Wang, W. G. Cheng. ***Tetrahedron Lett***., 2012, 53: 2684-2688; b) J. Sun, L. J. Han, W. G. Cheng , J. Q. Wang, X. P. Zhang, S. J. Zhang. ***ChemSusChem.***, 2011, 4(4): 502-507; c) J. Sun, S. J. Zhang, W. G. Cheng, J. Y. Ren. ***Tetrahedron Lett.***, 2008, 49(22): 3588-3591.

[5] a) W. G. Cheng, X. Cheng, J. Sun, J.Q. Wang, S. J. Zhang. ***Catal. Today***, 2013, 200:117-124; b) J. Sun, J.Q. Wang, W. G. Cheng, J. X. Zhang, X. H. Li, S. J. Zhang, Y. B. She. ***Green Chem***., 2012, 14: 654-660. c) J. Sun, W. G. Cheng, W. Fan, Y. H. Wang, Z. Y. Meng, S. J. Zhang. ***Catal. Today***, 2009, 148: 361-367.

Acknowledgments

This work was supported by National Basic Research Program of China (2009CB219901), and National Science Fund of China (21006117).

(707c) Modeling of CO_2 Capture Using Microalgae

Andrade, L. M., University of Sao Paulo
Mendes, M. A., University of Sao Paulo
Silveira, C. A.,
Oller Nascimento, C. A., University of Sao Paulo
Originally presented on: 11/7/2013 13:20:00 - 13:40:00

In order to model the growth of microalgae in a open-pond reactor, a 1000L system has been built and operated under realistic outdoor conditions. Environmental data (light intensity, temperature, pH) were captured with a datalogger and the growth was measured. A phenomenological mathematical model was built for the open reactor. The comparison of the simulated data to the experimental data shows good agreement.

(707d) Biogas Sweetening Via a Carbamate Crystallization Approach

Zheng, S., ZhejiangUniversity
Liu, Q.,
He, Y.,
Shi, Y.,
Originally presented on: 11/7/2013 13:40:00 - 14:00:00

The CO_2 in biogas significantly limited its potential applications. The regeneration process of aqueous alkanolamine based CO_2 capture often requires a large amount of energy, which makes the process costly. In order to enhance the energy efficiency of the capture, a carbamate crystallization approach was proposed in this study where water was replaced with non-aqueous solvents. It was observed that triethylenetetramine (TETA) carbamate crystallize out as supernatant solid from ethanol solution upon the adsorption of CO_2. A TETA/non-aqueous solvent solution shows higher absorption capacity and absorption rate than alkanolamine-water solution. The mole ratio of CO_2/TETA in carbamate was determined to be around 2:1 by elemental analyzer. The regeneration capabilities of solid carbamates was also investigated. Thermal behavior of the carbamate was examined with differential scanning calorimetry and thermogravimetric analysis. It was shown that the decomposition of TETA-carbamate commenced at about 90°C. The separation of carbamate as solid phase can offer great advantage by letting less volume to regenerate and make this approach a promising alternative for biogas sweetening.

(707e) Preparation and Absorption Behaviors Study of Doped Li4SiO4 At Low CO2 Concentration

Qi, Z., East China University of Science and Technology
Sai, Z., East China University of Science and Technology
Originally presented on: 11/7/2013 14:00:00 - 14:20:00

Preparation and Absorption Behaviors Study of Doped Li_4SiO_4 at Low CO_2 Concentration

Sai Zhang, Qi Zhang*,Qian Ye, Haiyan Wang, Zibin Zhu

Dept. of Chemical Engineering, East China University of Science and Technology, Shanghai 200237, China

Abstract

Different K-doped Li_4SiO_4 samples were prepared to improve the pure Li_4SiO_4 adsorbent's sorption/desorption properties in the critical SESMR practical conditions with low CO_2 partial pressure and steam atmosphere. The optimized sample was analyzed by sorption experiments and simulation with FactSage5.5 Software. The results show that the K-doped Li_4SiO_4 has a better CO_2 absorption ability than pure Li_4SiO_4 as the K can form a eutectic compound with Li_4SiO_4 which enhances the Li^+ and O^{2-} ions diffusion and thus facilitating the CO_2 absorption ability of Li_4SiO_4. It is also found that steam not only enhances the Li^+ mobility but also reacts with Li_4SiO_4, therefore increase the mass uptake of the adsorbent. In order to make the adsorbent more practical, sorption-desorption cycle processes were also carried out. The results show that the optimized Li_4SiO_4 has a lower desorption temperature and stable absorption ability.

Furthermore, the sorbents, Li_4SiO_4 and different K-doped Li_4SiO_4, were evaluated using a bench scale unit, of which the results can provide reference for the subsequent industrial design. Li_4SiO_4 and different K-doped Li_4SiO_4 were prepared by solid-state reaction method. The results was obtained from the studies of different Li_4SiO_4 samples in a fixed-bed reactor with 15%CO_2 and 85%N_2 at 575Â°Ã¦. The pure Li_4SiO_4 showed a short breakthrough time, approximately 20 min, while modified Li_4SiO_4 samples presented a relative 20~40 min longer breakthrough time under same conditions. The results of the experiment indicate that doping K into Li_4SiO_4 can improve the adsorption capacity of the sorbents. Comparing with MgO-based K_2CO_3[1] and hydrotalcites[2] sorbents, which breakthrough time are no more than 40min and 13min, respectively, the relatively long breakthrough time of K-doped Li_4SiO_4 exhibits huge potential for industrial applications. K-Li_4SiO_4 with good performance at low CO_2 partial pressure can be applied into sorption enhanced steam methane reforming system, in which the CO_2 concentration can be also as low as 15%. Meanwhile, the regeneration period can be prolonged by doping appropriate amount K into Li4SiO4, which means cost for recycling can be reduced as a result. K-doped Li4SiO4 has a certain value in industrial application, both practicability and economically.

Key words: CO_2 acceptors, Lithium orthosilicate, on-line absorption behaviors

References

[1]Soo Chool Lee, Ho Jin Chae, et al. Development of Regenerable MgO-Based Sorbent Promoted with K2CO3 for CO2 Capture at Low Temperatures[J]. Environ.Sci. Technol, 2008, 42: 2736-2741

[2]Hendricus Th. J. Reijers, Sacha E. A.Valster-schiermeier, et al. Hydrotalcite as CO2 Sorbent for Sorption-Enhanced Steam Reforming of Methane[J]. Ind. Eng. Chem. Res, 2006, 45, 2522-2530

(707f) Zr Doping Effect On CeO2 As the Catalyst in the CO2 - Ethylbenzene Reaction

Wang, Q.,
Li, W., Taiyuan University of Technology
Originally presented on: 11/7/2013 14:20:00 - 14:40:00

Under the thermodynamic limitation , dehydrogenation of ethylbenzene (EB) to styrene (ST) with CO_2 is widely investigated. The lattice oxygen mobile capacity of catalyst is the important influencing factors in the CO_2-EB reaction. CeO_2 of the CaF_2 structure with the excellent oxygen storage/release capacity (OSC) is preferred as the catalyst. The addition of Zr^{4+} to the cubic structure of CeO_2 can greatly increase the OSC and improve its surface area, thermal stability, mechanical strength and surface acidity. The Zr content of $Ce_xZr_{1-x}O_2$ solid solutions strongly influence the crystal structure, the mobility of lattice oxygen and the catalytic performance. $Ce_xZr_{1-x}O_2$ solid solutions replacing the pure CeO_2 as the catalyst could be used in oxidative dehydrogenation of EB to ST with CO_2 for changing the lattice structure of CeO_2 and improving the transmission capacity of lattice oxygen.

In this work, $Ce_xZr_{1-x}O_2$ solid solutions with different Zr proportions were synthesized using hydrothermal method. The synthesized catalysts were characterized by X-ray power diffraction pattern, Raman spectroscopy and N_2-adsorption. It was found that the catalysts were solid solution mixture having specific surface area more than 100 m^2/g. By the means of the H_2 temperature-programmed reduction, the lattice oxygen mobile capacity was calculated, corresponding to the order is $Ce_{0.5}Zr_{0.5}O_2 > Ce_{0.7}Zr_{0.3}O_2 > Ce_{0.3}Zr_{0.7}O_2 > CeO_2$. The performance of the catalysts for dehydrogenation of EB to ST was evaluated in a fixed-bed reactor under the conditions 823 K, 1 atm, and 20 of the CO_2/EB molar ratio. Results showed that the lattice oxygen mobile capacity of $Ce_xZr_{1-x}O_2$ influenced the catalytic activity and promoted activity of CO_2. The $Ce_{0.5}Zr_{0.5}O_2$ showed the highest activity with the EB conversion of 50% and the ST selectivity above 96%. The mechanistic model of EB dehydrogenation with CO_2 over the $Ce_xZr_{1-x}O_2$ catalysts was proposed.

(707g) Combined Natural Gas Liquefaction and Power Generation Plant With Carbon Dioxide Sequestration

Verma, S., Schlumberger Doll Research
Ramakrishnan, T. S.,
Originally presented on: 11/7/2013 14:40:00 - 15:00:00

It is generally well accepted that an increase in anthropogenic emissions is a leading cause of the increase in levels of carbon dioxide (CO_2), one of the greenhouse gases, in the atmosphere. Generation of electrical power is responsible for a roughly 40% of these emissions. Design of new zero emission power plants is critical in any effort to reduce emissions of CO_2.

We discuss the design of a combined natural gas liquefaction and power generation plant with no atmospheric emissions of greenhouse gases. More specifically, this design relates to (i) a zero emission process for power generation as the CO_2produced during the power generation cycle is sequestered into a terrestrial formation and (ii) a process by which the high pressure natural gas stream is expanded to provide refrigeration and power. The recovered energy may be used to liquefy part of the natural gas stream to provide pure liquefied natural gas (LNG).

This plant uses natural gas from a well as a fuel. The high-pressure stream is liquefied in a reverse-Brayton cycle and the remainder of the natural gas stream is fed to a solid oxide fuel cell to produce power. The effluent of the solid oxide fuel cell is completely combusted to reduce it to CO_2 and water. This stream is subsequently cooled to remove the condensed water.

Energy recovered in expansion of natural gas is used for CO2 stream compression, reducing the energy required to by 13.7%. Alternatively, if LNG is not desired as a product, then as much as 30% of the compression energy may be supplied from expansion of natural gas. Overall power plant efficiencies of 50 – 60% are possible. About 27.5% wt/wt of the incoming natural gas stream may be liquefied in this cycle.

(709a) Computational Study of the Proton Affinity, 1 H NMR Chemical Shifts, and Thermophysical Properties of Imidazole Derivatives

Liu, H., University of Alabama
Bara, J. E., University of Alabama
Turner, C. H., University of Alabama
Originally presented on: 11/7/2013 15:15:00 - 15:36:00

There has been increasing interest in developing new materials for CO_2 capture. One of the approaches is to synthesize polymer membrane using imidazole derivatives as the monomer. These imidazole derivatives are able to react with aqueous CO_2 to form bicarbonate anions, thus enabling facilitated transport of CO_2 through the membrane. In order to better understand the chemical and thermophysical properties of imidazole derivatives, we have employed quantum mechanics and molecular dynamics methods to calculate thc proton affinitics, ^{1}H NMR chemical shifts, and absorption properties of a variety of imidazole derivatives. From our simulations, a quantitative characterization of the effect of various exocyclic substituents on the proton affinity of imidazole was achieved. In particular, we find that electron-donating groups are able to increase the proton affinity of imidazole. In addition, the calculated ^{1}H NMR chemical shift and thermophysical properties are in excellent with experiment. These results will be helpful for future experimental design of new materials to be used for CO_2 capture.

(709b) Separation Selectivity of Various Gases in the Ionic Liquid 1-Ethyl-3 -Methylimidazolium Tris(pentafluoroethyl)Trifluorophosphate

Althuluth, M., The Petroleum Institute
Peters, C. J., Petroleum Institute
Berrouk, A. S., Petroleum Institute
Kroon, M. C., Eindhoven University of Technology
Originally presented on: 11/7/2013 15:36:00 - 15:57:00

The separation of carbon dioxide (CO_2) from methane (CH_4) is an important process in many industrial areas such as natural gas processing and biogas purification [1]. Natural gas also contains significant amounts of ethane, some propane, butane, and other higher hydrocarbons [2]. In natural gas treating, loss of heavy hydrocarbons is a concern. It is desirable to recover these compounds due to practical problems.

There are many methods available for the removal of acid gases from gas streams. The most commonly used are chemical solvents, physical solvents, membranes and cryogenic fractionation [3]. Physical solvents tend to be favoured over chemical solvents when the concentration of acid gases is very high. However, if the concentration of heavy hydrocarbons is high, a physical solvent may not be the best option due to higher co-absorption of hydrocarbons. An acceptable solvent should have a high capacity for acid gas and a low capacity for hydrocarbons [4].

Therefore, ILs have been proposed to be used as a solvent for CO_2 capture, because of their advantageous properties over conventional ones [5]. We found that the ionic liquid (IL) 1-ethyl-3 methylimidazoliumtris(pentafluoroethyl) trifluorophosphate ([emim][FAP]) shows the highest carbon dioxide (CO_2) solubility of all ILs studied so far and shows higher selectivities for CO_2/CH_4 separations than any other IL, indicating the promising potential of using this IL for the separation of CO_2 from natural gas. Also, the solubilities of C_2H_6, C_3H_8 and C_4H_{10} in ([emim][FAP]) were measured and compared to the CO_2 solubility in the same IL.

The separation ratio between CO_2 and hydrocarbons decreases as the hydrocarbon chain becomes longer. The selectivity increase accordingly to the order: $S_{CO2/C4H10} < S_{CO2/C3H8} < S_{CO2/C2H6} < S_{CO2/CH4}$. A maximum selectivity is achieved at lower temperatures. Therefore, the CO_2 removal from a natural gas stream is recommended to be performed at low temperatures in order to achieve the best separation.

The obtained solubility data are modelled with the Peng-Robinson equation of state combined with quadratic mixing rule. The calculated data have been found to be in a good agreement with the experimental results.

Refrences:

[1] M. Arruebo, J. Coronas, M. MenÃ©ndez, J. Santamaría, Separation and Purification Technology 25 (2001) 275-286.

[1] R.W. Baker, K. Lokhandwala, Industrial & Engineering Chemistry Research 47 (2008) 2109-2121.

[3] G. Aldana, R. Arai, D. G. Elliot, An Evaluation of Sources of CO2 for EOR in Venezuela, Proceedings of the Sixty-Third Annual Gas Processors Association Convention, New Orleans, Louisiana, 1984.

[4] N. Korens, D. R. Simbeck, D. J. Wilhelm, Process Screening Analysis of Alternative Gas Treating and Sulfur Removal for Gasification, Prepared for U.S. Department of Energy by SFA Pacific, Inc., Revised Final Report, December 2002.

[5] E.D. Bates, R.D. Mayton, I. Ntai, J.H. Davis, Journal of the American Chemical Society 124 (2002) 926-927.

(709c) Modelling and Measuring Hydrate Promotion for CO2 Capture

Herslund, P. J.,
Thomsen, K., Technical University of Denmark
Abildskov, J., Technical University of Denmark
von Solms, N., Technical University of Denmark (DTU)
GalfrÃ©, A.,
BrÃ¢ntuas, P., Ecole Nationale SupÃ©rieure des Mines de St-Etienne
Kwaterski, M.,
Herri, J. M.,
Originally presented on: 11/7/2013 15:57:00 - 16:18:00

A novel separation concept is investigated here, whereby a precipitated, or slurry, system, enables CO_2 capture by hydrate formation – enhanced by thermodynamic promoters. We have demonstrated that a model for the hydrate phase, combined with an equation of state for associating systems – the Cubic Plus Association (CPA) equation of state enables an accurate determination of mixed hydrate dissociation pressures as well as hydrate composition.

We have subsequently presented experimental work measuring promoted gas hydrate formation with tetrahydrofuran (THF) and cyclopentane (CP), both known low-pressure/high temperature sII hydrate formers. However when applied as a mixed promoter system the hydrate formation pressures are lowered by up to 20% when compared to the pure CP promoter system, which is thermodynamically the more efficient of the two pure promoter systems.

Finally, in order to model this promoted system the complex phase behaviour between water and THF needs to be understood. This system is successfully modeled using CPA, including the complex LLE observed at high pressures. This complex system is best modeled by specifically accounting for the solvation of THF (i.e. that THF forms hydrogen bonds with water but is not self-associating), which is also the correct physical picture of what occurs.

(709d) Experimental Characterization Of Multiply-Substituted Alkylimidazoles As Solvents and Polymeric Membranes For CO2 Capture

Bara, J. E., University of Alabama
Shannon, M. S., University of Alabama
Moon, J., University of Alabama
Reclusado, K., Villanova University
Originally presented on: 11/7/2013 16:18:00 - 16:39:00

Previous work has characterized specific physical properties and CO_2 solubility of 1-*n*-alkylimidazole solvents for use in pre- and post-combustion CO_2 capture during gas treating. This research project explores the CO_2 solubility and physical properties of multiply-branched alkylimidazole solvents to determine the effects of functional group position(s) on various chemical and physical attributes. Using this information, additional correlations between ring substitution and physical properties can be derived to optimize the characteristics best suited for larger-scale pilot testing in industrial applications. Synthesis and purification of these compounds were followed by tests to determine density, viscosity, and vapor pressure as functions of temperature and ring substitution, and these properties were shown to be comparable to previously-tested 1-*n*-alkylimidazoles. The solvents in general were found to have low viscosities (<15 cP) and negligible vapor pressures (<0.5 mbar). CO_2 solubility experiments also showed that multiply-branched alkylimidazoles have similar if not improved absorption capacities relative to ionic liquids and 1-*n*-alkylimidazoles. In future work, polymeric membranes can be synthesized from these compounds to determine CO_2 permeability and selectivity as a function of ring substitution. The ultimate goal of this research is to find an imidazole-based solvent with low volatility and high CO_2 absorption capacity or an imidazole-based membrane structure with high CO_2 permeability and selectivity that can be easily manufactured for application in large-scale industry for the reduction of atmospheric CO_2 emissions.

(709e) Tuning the Physical and Chemical Properties of Novel CO2 Capture Solvents Based On Hybrid Nanomaterials

Petit, C., Imperial College London
Park, A. H. A., Columbia University
Originally presented on: 11/7/2013 16:39:00 - 17:00:00

Nanoparticle Organic Hybrid Materials (NOHMs) represent a novel class of liquid-like materials, whose potential applications include CO_2 capture. Their negligible vapor pressure and high thermal stability have indeed prompted the research community to investigate these nanomaterials as alternative carbon capture media to the common amine-based solvents. NOHMs are synthesized by grafting a nanosized inorganic core with a polymeric canopy. Prior studies have shed light on the mechanisms of CO_2 capture in NOHMs as well as their capture capacity and selectivity. In this study, the investigation of CO_2 capture using NOHMs is continued to investigate the effects of NOHMs' chemistry (*e.g.*, effect of amines functionalities) and structure (*e.g.*, effect of core size) on tuning the kinetic (*e.g.*, CO_2 diffusion), thermodynamic (*e.g.*, heat capacity), absorptive (*e.g.*, selectivity compared to H_2O and SO_2) and physical (*e.g.*, thermal stability) properties of NOHMs in terms of CO_2 capture. A range of analytical techniques are employed including various spectroscopical techniques (*i.e.*, ATR FT-IR, NMR, Raman), thermogravimetry and rheometry.

(709f) The Study of Using Novel Scrubbing Agent Mg(OH)2 Solution to Capture CO2 From Flue Gases

Cheng, L., University of Cincinnati
Li, T., University of cincinnati
Keener, T. C., University of Cincinnati
Lee, J. Y., University of Cincinnati
Originally presented on: 11/7/2013 17:00:00 - 17:21:00

Abstract

A novel CO_2 capture method by using environment-friendly chemical magnesium hydroxide solution has been tested and evaluated in response to minimizing the CO_2 emissions from coal-fired power plant in the post-combustion control area. The goal of this study is to conduct the technical feasibility studies in the area of physical &chemical characteristic of $Mg(OH)_2$, chemistry of the system, lean solution performance in absorption, capacity of regeneration in desorption. It is found that >90% removal efficiency can be achieved in the bubble column absorber; a C/Mg ratio of 9 is achievable at the condition of 52 ℃absorption and 75 ℃ desorption; a fresh feeding rate of 0.008 L/sec is optimized condition in the continuous test.

(709g) Electrochemical Strategies for Reduction of Greenhouse Gas Emissions

Stern, M. C., Massachusetts Institute of Technology
Herzog, H., Massachusetts Institute of Technology
Hatton, T. A., Massachusetts Institute of Technology
Eltayeb, A.,

Originally presented on: 11/7/2013 17:21:00 - 17:42:00

Anthropogenic carbon dioxide (CO_2) in the Earth's atmosphere has been cited as a primary cause of global climate change and threatens global public health and welfare. Carbon Capture and Sequestration (CCS) is an effective and important part of CO_2 emission abatement strategies, with the major CCS efforts to date focusing on the removal of directly from large-scale carbon emitters and storing it in secure geologic reservoirs. Thermal-swing operations using aqueous base scrubbing followed by stripping at elevated temperature have been the chemical sorption processes most investigated over the past two decades for CO_2 capture. Considerable quantities of steam and heat are required to release the CO_2 after capture at low temperatures, and substantial parasitic energy losses result from the need to use excess steam and heat in order to meet the kinetic requirements of the process.

Electrochemically-mediated amine regeneration is a new post-combustion capture technology with the potential to exploit the excellent removal efficiencies of thermal amine scrubbers while reducing parasitic energy losses and capital costs. The improvements result from the use of an electrochemical stripping cycle, in lieu of the traditional thermal swing, to facilitate CO_2 desorption and amine regeneration; metal cations generated at an anode react with the amines, displacing the CO_2, which is then flashed off, and the amines are regenerated by subsequent reduction of the metal cations in a cathode cell. The advantages of such a process include higher CO_2 desorption pressures, smaller absorbers, and lower energy demands. Several example chemistries using different polyamines and copper are presented. Experimental results indicate an open-circuit efficiency of 54% (15 kJ/mole CO_2) is achievable at the tested conditions and models predict that 69% efficiency is possible at higher temperatures and pressures. A bench scale system produced 1.6 mL/min CO_2 while operating at 0.4 volts and 42% Faradaic efficiency; this corresponds to a work of less than 100 kJ/mole.

(713a) Integrated Absorber-Regenerator System for Rapid Screening of Thermal and Thermo-Oxidative Degradation of Primary and Secondary Amines for Predicting Fate in Validation-Scale CO2 Capture (CC) Systems

Jayaweera, I.,
Jayaweera, P.,
Bao, J.,
Bhamidi, S., SRI International
Originally presented on: 11/7/2013 15:15:00 - 15:40:00

Primary and secondary alkanoamine-based systems are currently considered the technology of choice for early demonstration of CO_2 capture (CC) solutions for power plants. Therefore, it is of great interest to predict the fate of amines over long periods of cyclic operations. We have incorporated existing thermal decomposition data on MEA (monoethanol amine) to design and operate an integrated absorber-regenerator system for rapid screening and simulation of the cyclic absorption of CO_2 and regeneration of the amines. The integrated rapid screening system includes: (1) the absorber; (2) the regenerator, and (3) the online electrochemical monitoring system for assessing the corrosion potential of stable and transient species from the amine degradation. The screening process is validated by matching the historic rate data for thermal and thermo-oxidative degradation of MEA in the temperature 50 to 175°C with varying parametric conditions (amine concentration, O_2 concentration and CO_2 loading). The data from the screening of varying primary and secondary amine compositions with conditions relevant to validation-scale CC systems will be presented.

(713c) Development of Fast and Thermally Stable Solvent System for CO2 Capture From Post-Combustion Flue Gas

Liu, K., Center for Applied Energy Research, University of Kentucky
Matin, N. S., Center for Applied Energy Research, University of Kentucky
Huang, Q., Center for Applied Energy Research, University of Kentucky
Martin, C., Center for Applied Energy Research, University of Kentucky
Remias, J. E., University of Kentucky
Liu, K., University of Kentucky
Originally presented on: 11/7/2013 16:05:00 - 16:30:00

Aqueous amine solutions have been identified as one of the most promising solvent systems for CO_2 capture from coal combustion. Despite the advantages of for example the well-known solvent system, such as aqueous monoethanolamine (MEA), the cost to capture CO_2 from coal combustion flue gas using such solvent system is expected to increase the electricity cost by 30% - 40%. The development of the novel solvent (CAER-S2) in the Center for Applied Energy Center at the University of Kentucky could largely reduce the operating and capital cost for CO_2 capture. The CAER-S2 solvent system has the advantages of fast CO_2 removal rate, high carbon cyclic loading, high thermal stability, low volatility, and commercial availability. Tests on both solvent screening apparatus and wetted-wall column at simulated scrubber condition indicate that the CO_2 removal rate of CAER-S2 solvent is more than 20% greater than 30 wt% MEA in the working carbon loading range. Such high removal rate helps reduce the absorber size. Vapor liquid equilibrium was also collected and discussed in terms of the working carbon loading range and heat of absorption. Thermal stability study performed at 125 °C and 135 °C shows that CAER-S2 can be operated at about 10 °C higher than that of 30 wt% MEA at given similar degradation rate. Operating at elevated temperature increases the CO_2 partial pressure at the top of stripper, lowering process energy by reducing the steam to CO_2 ratio and auxiliary energy for CO_2 compression.

(713d) Advanced Stripper Configurations for CO2 Capture Using PZ and MEA

Lin, Y. J., University of Texas at Austin
Madan, T., The University of Texas at Austin
Rochelle, G. T., The University of Texas at Austin
Originally presented on: 11/7/2013 16:30:00 - 16:55:00

For post-combustion CO_2 capture, steam usage for lean solvent regeneration in the stripper and CO_2 compression work are the main contributions to the energy requirement of the process. Implementing CO_2 capture incurs a 20–30% penalty of electricity output for a typical coal-fired power plant. Alternative stripper configurations could improve energy efficiency significantly compared to a simple stripper. Loss of steam heat from the stripper is one of the reasons that CO_2 capture by amine scrubbing is inefficient. When the stripper is operated at 120–150 °C, water in the rich solvent is vaporized and emitted with CO_2 from the top of stripper. The relative amount of water vaporized to CO_2 removed depends on solvent properties. Solvents with different heat of absorption and tolerable operating temperature preventing dramatic thermal degradation decide how much stripping steam can be recovered. In this work, PZ and MEA were both investigated.

The objective of this work is to investigate the improvement of rich exchanger bypass strategy applying a heat exchanger to recover steam heat from CO_2 vapor by bypassing cold rich solvent. Several advanced stripper configurations have been modeled and optimized using Aspen Plus®. Equivalent work is used as an indicator of energy performance instead of only heat duty. Rich exchanger bypass is applied to advanced stripper configurations including interheated stripper, stripper with warm rich bypass and flash stripper with warm rich bypass. Also, 9 m MEA and 8 m PZ are both investigated.

Compared to a simple stripper, the rich exchanger bypass configuration has 7.2% improvement of equivalent work for PZ and 3.7% for MEA. The interheated stripper and stripper with warm rich bypass have similar energy performance with the lowest equivalent work. The interheated stripper integrated with rich exchanger bypass has 8.7% improvement for PZ and 5.4% for MEA. The stripper with warm rich bypass and rich exchanger bypass has 9.2% improvement for PZ and 5.5% for MEA. A flash stripper with warm rich bypass and rich exchanger bypass configuration has 8.4% improvement for PZ and 4.4% for MEA. Because the flash stripper can use a smaller solvent residence time at elevated temperature, it can be designed to operate at greater temperature and pressure, with further energy savings.

(713e) Rigorous Modeling of Novel Absorption-Stripping Process Configurations With Aqueous Piperazine for CO2 Capture From Natural Gas Combustion

Sachde, D. J., The University of Texas at Austin

Rochelle, G. T., The University of Texas at Austin

Madan, T., The University of Texas at Austin

Originally presented on: 11/7/2013 16:55:00 - 17:20:00

This study evaluates amine scrubbing using 8 m piperazine (PZ) for CO2 removal from the flue gas of three natural gas-based applications. The applications and corresponding flue gas CO2 concentrations are as follows: Case 1, combined cycle (3% CO2), Case 2, combined cycle with exhaust gas recycle (6% CO2), and Case 3, natural gas-fired boiler (9% CO2). For each application, two bounding design scenarios were evaluated: high capital and low operating cost (high cap-ex, low op-ex) or low capital and high operating cost (low cap-ex, high op-ex). Capital and operating costs were not evaluated explicitly; rather, within the two major process areas (absorption and stripping), equipment level design modifications and process operating conditions were used to approximate the economic scenarios. Process and equipment models were developed in Aspen Plus® utilizing a rigorous thermodynamic and kinetic framework developed previously for the piperazine solvent. The focus of this work is the development and modeling of new process and equipment configurations for the natural gas applications.

Key developments in absorber design included solvent recycle configurations for intercooling in the absorber. The recycle design was used for two purposes in the natural gas applications. First, the water content in the entering flue gas of the combined cycle applications (Cases 1 and 2) was sufficiently low (< 7% water) that cooling of the gas to 40°C (absorber operating temperature) would not lead to condensation of water. This allowed the elimination of a separate direct contact cooler (DCC) upstream of the absorber in favor of a packed section at the bottom of the absorber with recycle of a high rate (~ 3 times the solvent feed rate) of cool, rich solvent to couple the function of cooling the inlet flue gas with absorption of CO_2.

In addition, the liquid-to-gas flow ratio (L/G, mole to mole basis) in the column is much lower than in the typical coal applications (~12-14% CO_2 in the flue gas) due to the high volume of flue gas treated. The resulting energy balance in the absorber leads to a maximum in column temperature (and corresponding equilibrium limited mass transfer) towards the top of the column. In other words, the heat generated by absorption of CO_2 in the solvent is carried by the gas to the top of the column. Figure 1 depicts a temperature profile for Case 1 as an illustration of the expected temperature behavior in the gas applications.

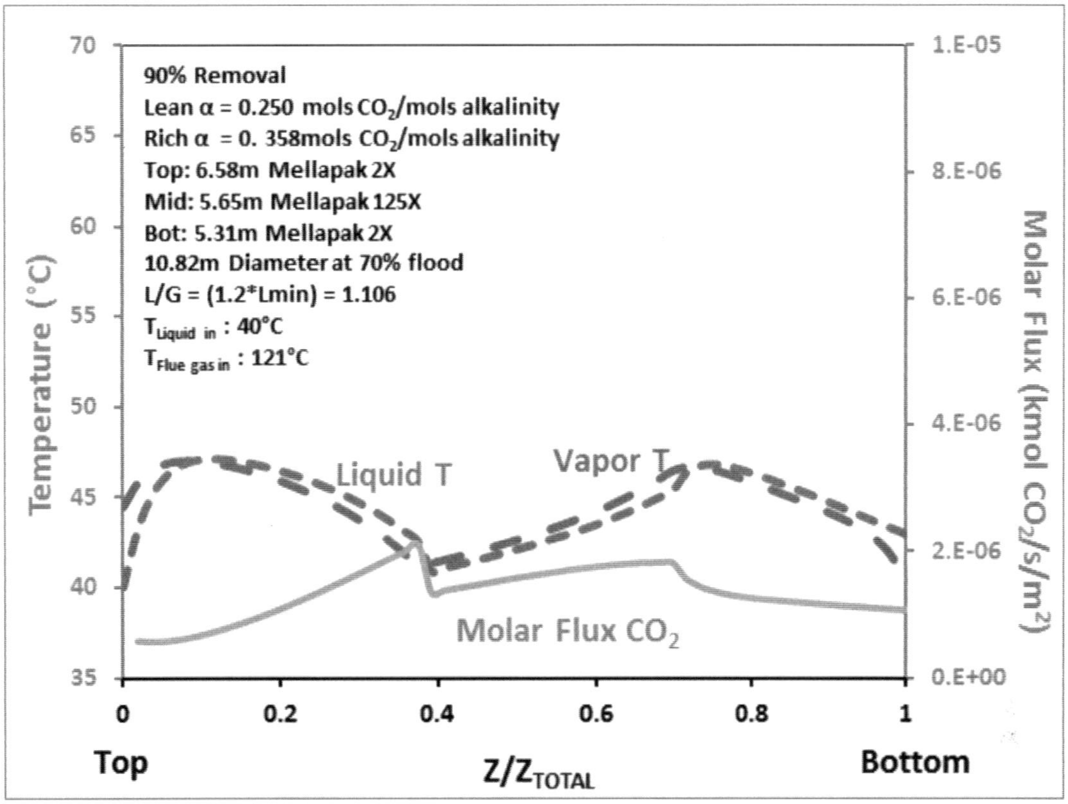

Figure
1: Molar Flux and temperature profiles, Case 1 (3%CO_2) with DCC. Lean loading = 0.25 mols CO_2/mols alkalinity, lean amine T = 40 °C, rich gas T = 40 °C, 3:1 recycle.

The solvent-recycle design depicted in Figure 2 is shown to cool the gas effectively by contacting it with a high rate of cooled solvent in a well-mixed section of the column, and, in turn, reducing the temperatures throughout the column. The high solvent rate per wetted perimeter also provides enhancement of mass transfer (via generation of turbulence in the recycle section). Sensitivity analyses identified solvent recycle rates to minimize total packing requirements (capital costs) and maximize rich loadings (minimize energy requirement in the stripping section).

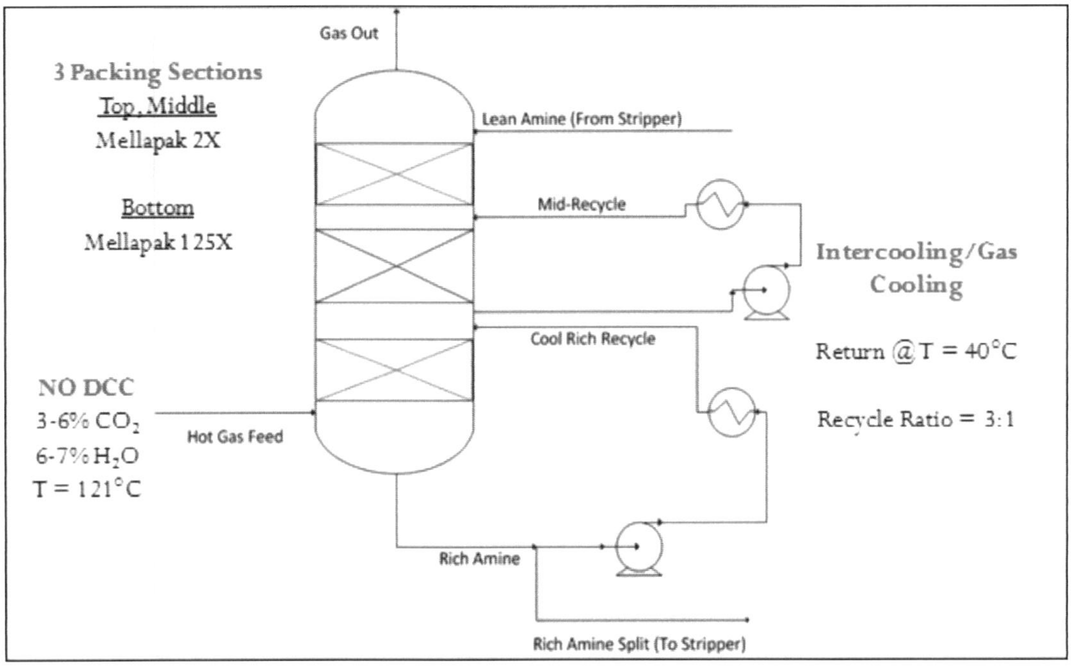

Figure 2: Case 1 and 2 Absorber PFD with Solvent Recycle, No DCC, Combined Cycle Gas Turbine.

In the stripping section, an inter-heated stripper was implemented with 5°C LMTD for the main cross exchanger (increased exchanger area) to represent a high cap-ex, low op-ex designs (Figure 3).

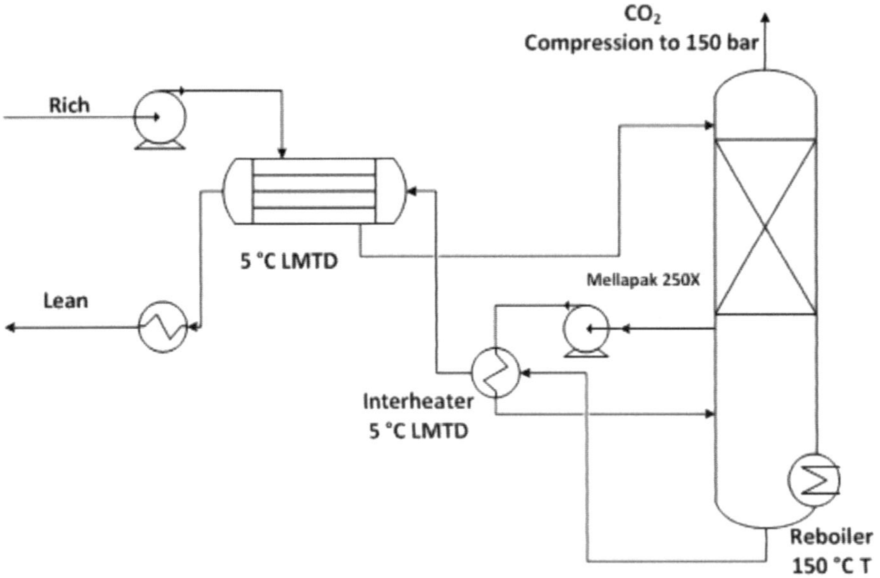

Figure 3: High cap-ex configuration of inter-heated stripper and 5 °C LMTD cross-exchanger

The inter-heated stripper configuration recovers additional heat from the reboiler via cross exchange of a solvent-draw from the stripper with the rich stream leaving the reboiler. The inter-heating also results in a cooler rich amine stream reaching the stripper since less heat is transferred in the main cross-exchanger (i.e., the heat was recovered in the inter-heater). The cooler rich amine stream entering the stripper reduces the amount of water vapor lost per mole of CO_2 recovered and in turn reduces irreversible heat losses in the condenser.

In addition, a simple stripper with cold rich bypass and 10°C LMTD for the main cross exchanger (reduced exchanger area) was implemented for the low cap-ex, high op-ex designs (Figure 4).

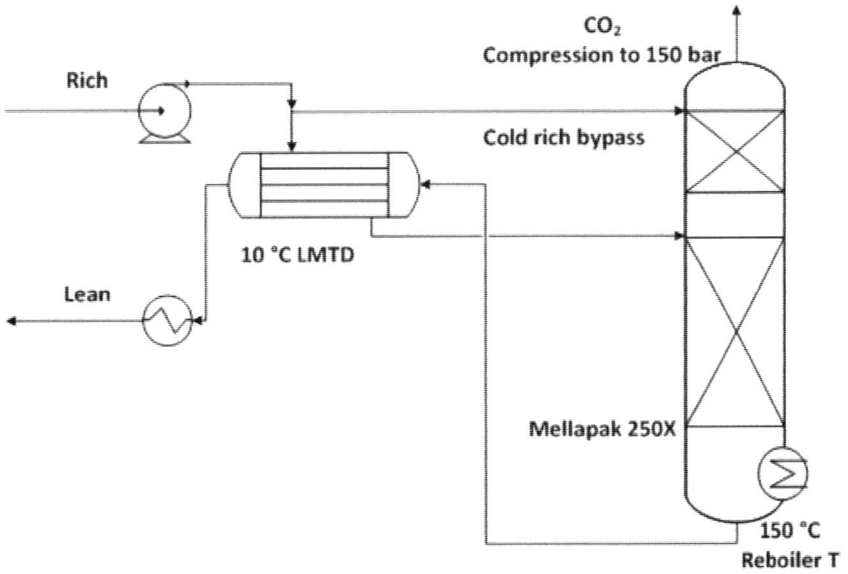

Figure 4: Low cap-ex configuration of simple stripper with cold rich bypass and 10 °C LMTD cross-exchanger

In this stripper design, the column consists of two packed sections. The cold rich amine stream leaving the absorber is split upstream of the main-cross exchanger. A fraction of the flow bypasses the main exchanger and is sent to the top of the stripper, reducing the water lost per mole of CO_2 in the gas leaving the stripper (as in the inter-heated stripper design). For both stripper configurations, analysis of the equivalent work requirement was conducted, including sensitivity analysis identifying a range of operable lean and rich loadings for the natural gas applications.

(713f) Integration of Solar Thermal Energy With Post-Combustion Carbon Capture Process

Parvareh, F., The University of Sydney
Sharma, M., The University of Sydney
Qadir, A., The University of Sydney
Khalilpour, R., The University of Sydney
Chiesa, M., Laboratory of Energy and Nano-Science (LENS)
Abbas, A., The University of Sydney

Originally presented on: 11/7/2013 17:20:00 - 17:45:00

Retrofitting existing conventional coal-fired power plants with solvent based post-combustion carbon capture (PCC) technology is a promising solution for CO2 emissions reductions. Such implementation is attractive because it sets a path of transition towards renewable energy, while continuing the use of fossil fuels in clean capture enabled energy production. The major drawback in PCC is its energy intensive solvent regeneration process. The required energy to regenerate the CO2 absorbing solvent is typically supplied by high quality steam, bled from the turbine circuit, which results in significant decrease in power plant efficiency. Integration of a solar thermal process with PCC-enabled power plants can improve the electricity production rate by totally or partially providing the thermal heat requirement for solvent regeneration.

In this paper, such an integrated system is modeled in dynamic mode to investigate transient process behavior as well as to identify the effects of solar energy input on overall system technical performance. Each subsystem (i.e. Power plant, PCC plant and Solar thermal plant) is modeled separately before integration of the three models within a model superstructure. This dynamic model superstructure is proposed for implementation and testing of optimal control policies, but significantly for economic optimization through integration with dynamic electricity and carbon market variables.

(769a) Evaluation of Reclaimer Sludge Disposal Options From Amine-Based Post-Combustion CO2 Capture Process

Sexton, A.,
Fisher, K.,
Chen, E., The University of Texas at Austin
Nielsen, P., The University of Texas at Austin
Dombrowski, K.,
Singh, P., IEAGHG
Originally presented on: 11/8/2013 8:30:00 - 8:55:00

An important environmental issue with respect to amine-based post-combustion CO_2 capture is the generation of amine waste containing degradation products and system impurities that must be mitigated or disposed of properly. The operation of an amine-based post-combustion CO_2 capture process will result in the formation of degradation products due to the presence of oxygen, NO_x and SO_x in the flue gas; additional impurities present in the flue gas (heavy metals, fly ash, halogen compounds) can be physically absorbed into the solvent and accumulate as it circulates within the CO_2capture system.

These degradation products exhibit corrosive properties and reduce solvent CO_2 absorption characteristics. To control the concentration of these degradation products in the amine- based solvent, a slipstream of the solvent from the CO_2 capture unit is sent to a solvent reclaiming system; the recovered amine-based solvent is returned to the CO_2capture unit, while waste from the reclaimer unit is discharged for disposal/treatment. This paper evaluates the environmental fate of reclaimer waste streams generated from three common amine-based solvents: monoethanolamine (MEA) , piperazine (PZ) and a methyldiethanolamine/piperazine blend (MDEA/PZ).

Reference cases will be developed for the supercritical pulverized coal (PC) and natural-gas combined cycle (NGCC) power plants at their respective gross power outputs, along with reference cases for the CO_2 capture processes integrated with the power facilities. These reference cases will identify the lean/rich loadings and circulation rate of the solvents of interest, in addition to the operating conditions of the regeneration column (used to characterize thermal degradation rates). Commentary will be provided on oxygen and trace impurities concentrations in the flue gas, and how these concentrations affect impurities and degradation product concentration within the circulating amine solvent adjusted to the overall CO_2capture rate. For other solvent systems (PZ, AMP/PZ, amino acids, etc.) where data are readily available, a qualitative analysis of degradation information will be provided.

For each solvent system, a degradation model will be developed to estimate solvent loss, impurities accumulation and degradation product accumulation expected in a commercial CO_2 capture unit. Solvent losses are expected via thermal degradation, oxidative degradation, amine volatility, reclaimer operation, nitrosamine formation from NO_x, and reactions with flue gas impurities. Thermal degradation losses will be modeled as a function of regeneration operating temperature and pressure, as well as rich loading of the amine-based solvent, while oxidative degradation losses will be modeled as a function of oxygen concentration, dissolved metal ion concentration and rich loading of the amine-based solvent. Reclaimer solvent losses will be modeled as a function of reclaimer inlet composition (includes degradation product and heavy metals concentrations) and reclaimer operating conditions. Solvent losses from flue gas impurities will be modelled as a function of impurities concentrations in the flue gas and the rate of physical absorption of the individual impurities.

A technology feasibility and economic analysis of various reclaiming technologies (e.g., thermal and vacuum distillation, ion exchange, and electrodialysis) will be conducted for each solvent system; the analysis will include process descriptions, process flow diagrams, steady-state mass balances, energy requirements, utility requirements, equipment lists, costs and projected solvent losses. Reported economic metrics will include the levelized cost of electricity (LCOE) and cost of CO_2emissions avoided; energy requirements for each reclaiming process will be normalized in terms of kW/gallon of solvent reclaimed.

Key assumptions within the analysis include the expected composition of reclaimer feed, typical amine recovery in a thermal reclaiming system, recovery of amine impurities and degradation products from the amine solvent, and the fraction of amine degradation that occurs during thermal reclaiming. These assumptions will be developed from information gathered from journals, conference proceedings, conversations with amine waste treating industry representatives, and conversations with reclaiming vendors. In addition, conversations with reclaiming vendors and amine waste treating industry representatives will be used to better understand operating experiences and operating preferences depending upon the composition of the inlet gas stream being treated and operating conditions within the amine treating system.

The composition of the reclaimer waste generated from each solvent via each type of reclaimer technology will be estimated and then classified based on US and EU environmental regulations for hazardous waste; in addition, cursory evaluations will be performed based on regulations in the UK, Canada and Australia. The evaluation will include extracting specifically applicable sections and sub-sections of pertinent regulations and discussing how they might apply to the reclaimer waste. Specifically, heavy metals will be evaluated, and concentrations that trigger hazardous waste regulations will be specifically identified.

This paper will also evaluate several potential options for disposing the reclaimer sludge waste, such as conventional incineration, co-firing in a power plant boiler, use as an SNCR reagent, treating in a conventional wastewater treatment plant, and use in a cement kiln. For each option, regulatory and technical feasibility, potential extra processing that would be required, and indication of the relative costs will be provided.

Based upon the tasks mentioned, recommendations will be developed to identify optimal operating conditions to minimize solvent loss and to identify the main components of the reclaimer waste expected from the specified flue gas composition and capture conditions. In addition, a review of existing and foreseeable issues of reclaiming with likely capture solvents will be conducted to provide direction for future research and development of other advanced reclaiming concepts and waste disposal.

(769b) Stack Sampling and Analytical Methods for Post-Combustion Carbon Capture Amines and Degradation Products

Shaw, S., EPRI
Rohr, A.,
Attalla, M., CSIRO
Hoff, K. A.,
Originally presented on: 11/8/2013 8:55:00 - 9:20:00

The use of amine solvents in post-combustion carbon capture (PCCC) pilot plants and testing studies has increased over the last several years, and subsequently a clear understanding of the potential human and environmental impacts of those solvents has become more crucial. One critical need that is made clear from review of current efforts to estimate emissions of the amine solvents and their degradation products is the need to standardize relevant stack sampling and analytical methods. Unfortunately, this issue is complicated by the varying chemical and physical properties of the several compound classes of interest, stage of method development, and difficulty of dealing with sampling constraints of power plant stacks. This led to the 2012 development of an international working group to collaboratively evaluate efficacy of methods that already have or could be applied in power plant flue gas, to improve characterization of sampling artifacts, to determine if standardized method performance metrics can be created, and to potentially design a method shoot-out test protocol. The goal is to summarize lessons learned and set the stage for future guidelines for more accurate determination of emissions of amines, related compounds, and their degradation products. Methods reviewed included both online (e.g. Fourier transform infrared spectroscopy) and offline (e.g. sorbent trapping and thermal desorbtion) analysis for both gas and particle phase compounds. An overview of the findings and recommendations from both working group efforts will be summarized in this presentation.

(769c) A Combined Hazard and Life Cycle Assessment of Six Amine Solvents for Post-Combustion CO2 Capture

Badr, S., Swiss Federal Institute of Technology, Zurich (ETHZ)
Frutiger, J., ETH Zurich
Tan, P. S., ETH Zurich
Papadokonstantakis, S., Swiss Federal Institute of Technology, Zurich (ETHZ)
Hungerbuehler, K., Swiss Federal Institute of Technology, Zurich (ETHZ)
Originally presented on: 11/8/2013 9:20:00 - 9:45:00

A multi-criteria assessment is carried out for 6 different solvents used for post-combustion CO2 capture. Monoethanolamine (MEA) is considered the standard solvent for CO2 capture. The performance of Methyldiethanolamine (MDEA), 2-Amino-2-methyl-1-propanol (AMP), Piperazine (PZ), Diethanolamine (DEA) and Diglycolamine (DGA) as solvents is compared to that of MEA from an Environmental, Health and Safety (EHS) hazard assessment and from a Life Cycle Assessment (LCA) perspective. Cumulative Energy Demand (CED), Global Warming Potential (GWP), Eco-Indicator 99 (EI99) and RECIPE are calculated for the capture process using each solvent. The EHS assessment covers the use of EHS method, the Inherent Safety Index (ISI) method, the Hazard Identification and Ranking (HIRA) method and the Integrated Inherent Safety Index (I2SI) method.

Solvent loss through degradation is accounted for in the assessment, as solvent replacement rates are included in the life cycle impact assessment and degradation products are included in the EHS hazard analysis. A framework for solvent degradation based on molecular structure is established based on experimental degradation rates available in literature. This framework then enables a more thorough assessment of more solvents even where experimental degradation data is scarce.

For the hazard analysis, a framework is established to overcome the issue of data gaps. In many cases and especially for the less used solvents, crucial properties required for the EHS assessment are missing such as (Threshold Limit Values) TLV and Immediately dangerous to Life and Health (IDLH) values. This makes it very difficult to use methods that only rely on such data without flexibility such as the ISI method. Hence the framework helps to give estimates for these values, translated from other properties, whenever they are missing.

The presented frameworks make it possible to extend the assessment to other solvents where little experimental data is available. DGA represents a case where TLV and degradation data are unavailable. Assessments are carried out with the application of both frameworks for estimating missing data.

(769d) Bicine Production in Promoted-Mdea Solvents

Closmann, F., Phillips 66

Originally presented on: 11/8/2013 9:45:00 - 10:10:00

The correlation between equipment corrosion and the presence of the amino acid bicine in acid gas scrubbing systems utilizing promoted methyldiethanolamine (MDEA) solvents is fairly well established. Preventing the formation of bicine in operating MDEA solvents is accepted as the best way to control corrosion. But the conditions conducive to bicine production are still being debated, with little consensus existing in the literature regarding the role of degradation product intermediates, dissolved oxygen, and H_2S scavengers (aldehydes) in bicine formation. Without a clear understanding of those conditions, bicine production will continue to be a problem for natural gas conditioning and refinery based amine solvent systems. This paper explores the conditions which lead to bicine production in MDEA-based solvents based on new laboratory results, and presents pathways which explain its formation.

Bicine production has been linked to the degradation intermediate product diethanolamine (DEA) in aqueous MDEA solvents (Closmann PhD dissertation, 2011) in the laboratory, and to triethanolamine (TEA) (Howard and Sargent, 2001) in field systems. MDEA-based solvents typically accumulate significant amounts of DEA, but little or no TEA. However TEA has been identified in MDEA degradation experiments, with its presence documented and explained mechanistically (Bedell, 2010). Bicine has been identified in separate oxidation experiments with DEA and TEA, leading to different explanations for its formation. Further, the role of two-carbon aldehydes in bicine production in degraded MDEA solvents is likely.

One explanation for the production of bicine in MDEA-based solvents comes from Closmann (2011), and involves the formation of DEA and other secondary amines through nucleophilic substitution reactions at the higher temperatures associated with the stripper reboiler (120+ °C). Once formed, the strong nucleophilic behavior of the secondary amine will lead to a reaction with carbonyl carbons associated with one and two-carbon aldehydes to form a substituted alcohol intermediate which can undergo rearrangement to form another aldehyde. This aldehyde will also behave as an intermediate, and will be readily oxidized to its respective amino acid. These two-carbon aldehyde interactions will lead to the formation of bicine. One such pathway for the formation of bicine through interaction between two reactive intermediates, DEA and hydroxyacetaldehyde, was proposed by Closmann (2011). The presence of both those intermediates can be explained in degraded MDEA solvents. The presence of other aldehydes (glyoxal) intentionally marketed as H_2S scavengers in incoming feed gas streams has been documented in field applications, and provides further support to this pathway for bicine production. In addition to the formation of secondary amines as intermediate products, our paper also explores these secondary amine-aldehyde interactions.

Another possible explanation for the presence of bicine in working solvents is the formation of TEA as an MDEA degradation intermediate. TEA, like other tertiary amines, could be formed through the same S_N2 substitution reactions serving to explain the formation of DEA and methylmonoethanolamine (MMEA) in degraded MDEA solvents. With the oxidation of a single hydroxyethyl group, TEA would be oxidized to bicine. We performed isolated oxidation experiments which readily demonstrated the formation of bicine from TEA. Those experiments were conducted in a jacketed glass reactor maintained at 55 °C, and fed a constant headspace gas of 98% oxygen at 100 cc/min. Dissolved transition metal sulfate salts of iron, chromium and nickel were added to catalyze oxidation activity in the aqueous solvent. The rate of bicine production in 40% (wt) TEA was 0.27 mmols/kg-hr, which was ~20X the rate observed in a 7 molal MDEA/2 molal PZ solvent oxidized at similar conditions in the same reactor. The formation of bicine in TEA was detected without a lag period, whereas a lag time of greater than 25 hours was observed in the promoted MDEA oxidation experiment, demonstrating the need to form and accumulate a reactive intermediate such as DEA in the promoted MDEA solvent. We will discuss the implications of a TEA-based pathway for bicine production in MDEA solvents.

We will also demonstrate that analogous pathways lead to the formation of other important amino acids in MDEA-based solvents. Those amino acids include hydroxyethyl sarcosine (HES) and N-(2-hydroxyethyl)glycine (HEG). We determined that the total carbon and nitrogen associated with these two amino acids is at least as much as that associated with bicine, and that the total nitrogen of all three of these amino acids can exceed 2% of all nitrogen in a working natural gas scrubbing solvent.

(769e) Thermal Decomposition of Nitrosamines in Amine Scrubbing

Fine, N. A., Washington University in St. Louis

Rochelle, G. T., The University of Texas at Austin

Originally presented on: 11/8/2013 10:10:00 - 10:35:00

Thermal Decomposition of Nitrosamines in Amine Scrubbing

Introduction

Unlike other carbon capture processes, amine scrubbing is a chemical absorption process which directly contacts the contaminated flue gas with a solvent. The contaminants in the flue gas can react with the solvent, forming dangerous byproducts under the high temperatures in the stripper. Nitrosamines, a family of organic compounds that form from secondary amines reacting with nitrosating agents, are one of the largest environmental issues associated with amine scrubbing. Over 80% of nitrosamines are carcinogenic to humans at low concentrations so it is important to understand and then control their accumulation in amine scrubbing. This work will describe thermal decomposition, the most straightforward techniques for controlling nitrosamine accumulation

Thermal Decomposition Kinetics

Recent research has shown that N-nitrosopiperazine (MNPZ) will decompose under stripper conditions (Fine & Rochelle, 2013). Decomposition is first order in MNPZ and half-order in the amine concentration. This work has been expanded to include the thermal decomposition of n-nitrosodiethanolamine (NDELA), nitrosopiperidine (NPIP), and nitrosomorpholine (NMOR) to understand how nitrosamine structure effects thermal degradation. Furthermore, the role of the solvent has been tested on both MNPZ and NDELA. Nitrosamine decomposition was found to be base catalyzed with a Bronsted slope of 0.5 (Figure 1). Base catalysis may be useful in amine scrubbing since reclaiming often runs under strong basic conditions. Nitrosamine decomposition was also found to be highly temperature dependent with an activation energy of approximately 100 kJ/mol across several nitrosamines. Thus, it is important to have a thermally stable solvent so that the stripper can run at higher temperature to increase nitrosamine decomposition.

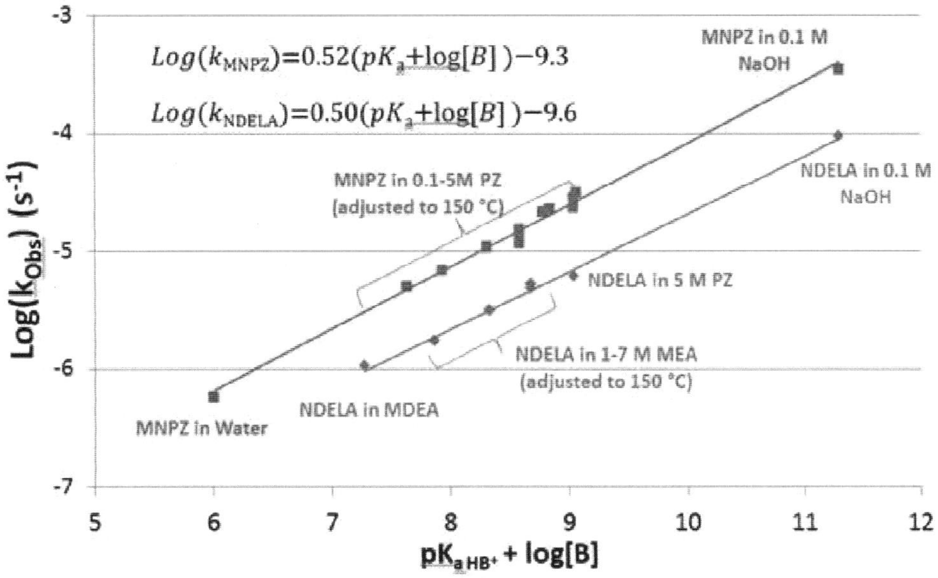

Figure 1: Base Catalysis in Nitrosamine Thermal Decomposition Thermal Decomposition Products

Samples of thermally decomposed MNPZ were analyzed using a previous HPLC method (Fine, Goldman,

Nielsen, & Rochelle, 2013) as well as a group analysis method for total nitrosamines. (Dreshcer & Frank, 1978; Walters & Smith, 1983). Both methods agreed within the precision of the instruments, proving that MNPZ does not transnitrosate into another nitrosamine (Figure 2). In another experiment the decomposed nitrosamine was analyzed for total aldehydes. Aldehyde concentration increased stoichiometrically as the nitrosamine decomposed; the final sample was analyzed using high resolution mass spectrometry and the aldehyde was determined to be piperazin-2-ol (Figure 3).

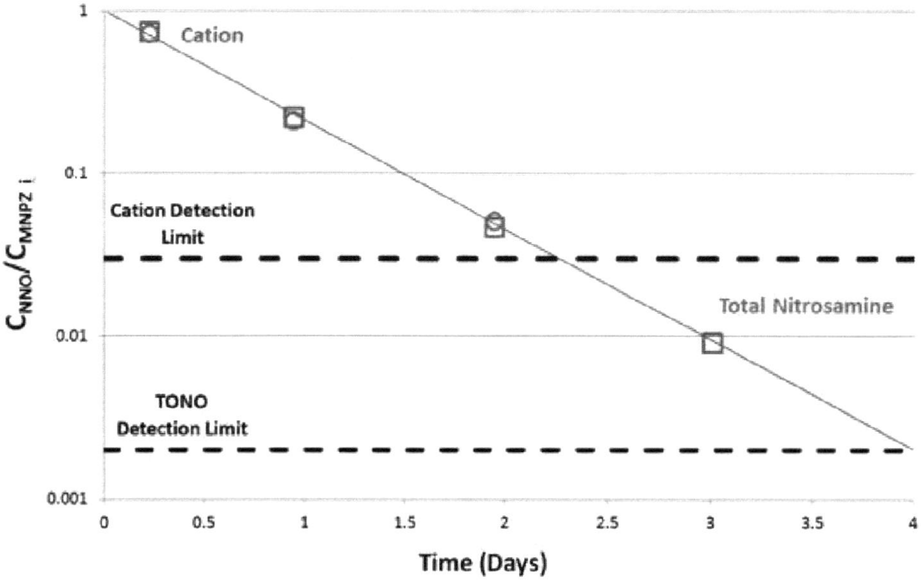

Figure 2: MNPZ Decomposition into Non-nitrosamine species

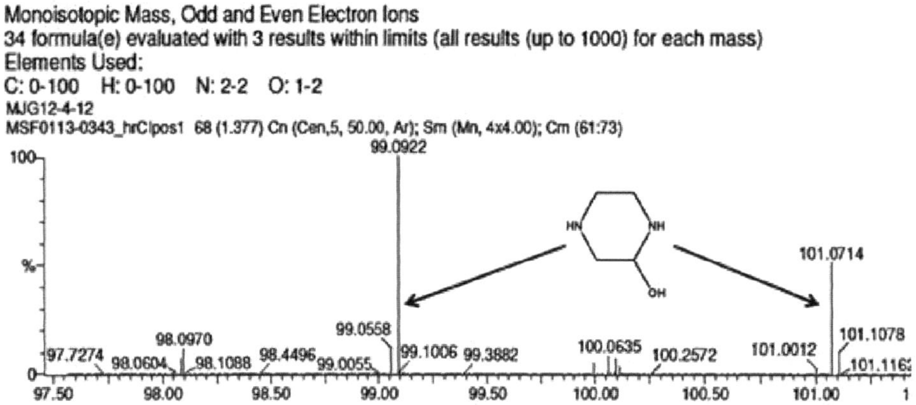

Minimum:	5.00				-1.5		
Maximum:	100.00		5.0	10.0	50.0		

Mass	RA	Calc. Mass	mDa	PPM	DBE	i-FIT	Formula
99.0558	12.04	99.0558	0.0	0.0	2.5	11.1	C4 H7 N2 O
100.0635	8.48	100.0637	-0.2	-2.0	2.0	76.1	C4 H8 N2 O
101.0714	51.41	101.0715	-0.1	-1.0	1.5	0.1	C4 H9 N2 O

High Resolution Mass Spectrometry of Piperazin-2-ol in Decomposed MNPZ Sample Thermal Decomposition Mechanism

We have now proposed a mechanism for nitrosamine thermal decomposition using the empirical kinetics and decomposition products found (Figure 4). The hypothesized formation of nitrous oxide will be studied using a cycling reactor. The piperazine imine form is readily hydrolyzed to form the piperazin-2-ol seen on the high resolution mass spectrometry.

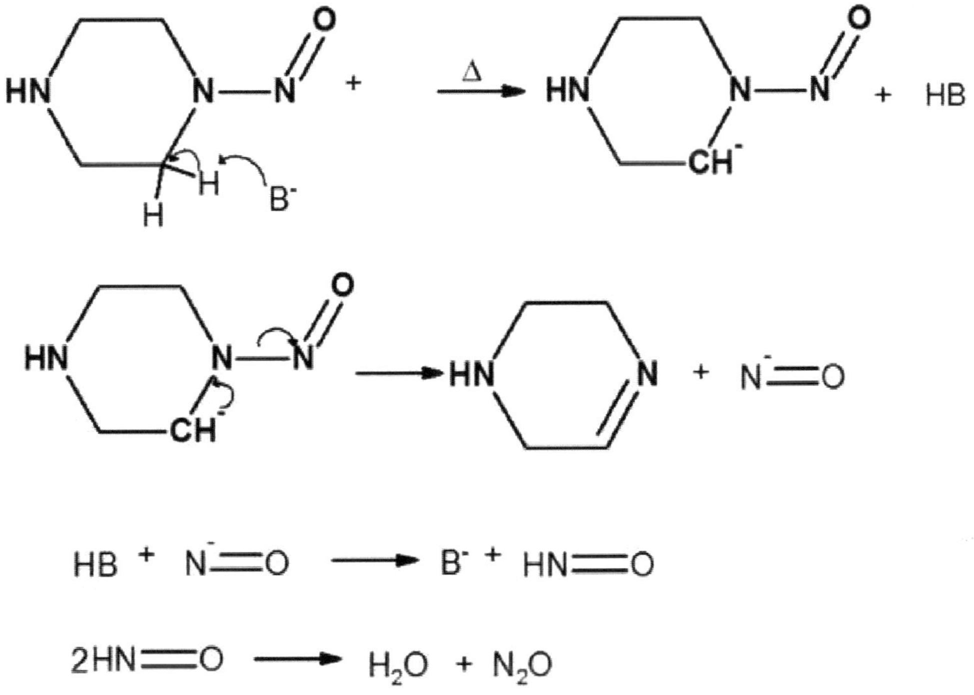

(769f) Influence of Amine Structural Characteristics On N-Nitrosamine Formation in Post-Combustion Carbon Dioxide Capture Systems

Dai, N., Yale University
Mitch, W. A., Yale University
Originally presented on: 11/8/2013 10:35:00 - 11:00:00

Amine-based CO_2 capture technology is the prime contender for implementing post-combustion CO_2 capture in the near future. Increasing attention has been paid to develop amines with structural characteristics featuring improved CO_2 absorption performance. However, the influence of amine structural characteristics on their environmental impacts has not been evaluated. Specifically, the formation of carcinogenic *N*-nitrosamines by NO_x reactions with amines within the capture unit and the possible contamination of downwind air and water by the release of these compounds are of great concern. In this study, 17 amines with systematically varied structural characteristics were selected and tested in a lab-scale CO_2 capture reactor for their *N*-nitrosamine formation potential. The order of the amine is a major factor affecting their total *N*-nitrosamine formation in the absorber unit. Primary amines formed total *N*-nitrosamines at 1-2 orders of magnitude lower rates compared to secondary and tertiary amines. The steric hindrance and alkyl chain length of primary amines had no significant impacts on *N*-nitrosamine formation in the absorber. When considering the total *N*-nitrosamine accumulation rates in the water wash unit, a parameter bearing direct environment relevance, amines with alcohol functional groups gave rise to lower total *N*-nitrosamine levels in the water wash unit, regardless of the order of the amino group. The influence of amine structural characteristics should be considered in the search for an "ideal" solvent for CO_2 capture to minimize environmental and health impacts related to *N*-nitrosamines.

(774a) A Novel Fluidized Bed Hydrator for Steam Reactivation of Calcium-Based CO2 Sorbent

Wang, A.,
Wang, D., The Ohio State University
Deshpande, N., The Ohio State University
Phalak, N., The Ohio State University
Wang, W., The Ohio State University
Fan, L. S., The Ohio State University
Originally presented on: 11/8/2013 8:30:00 - 8:55:00

Calcium-based solid sorbents has been widely applied for the removal of acid gases, such as sulfur dioxide (SO_2), sulfur trioxide (SO_3), and hydrochloric acid (HCl), from coal-fired flue gas. Amid possible carbon dioxide (CO_2) regulations, scientists have explored ways to extend this technology for CO_2 removal especially from sizable point sources at low cost. Here at the Ohio State University (OSU), researchers have developed a 3-step CaO-based carbonation-calcination reaction (CCR) process that operates in the range of 450 – 950 °C for the simultaneous removal of both CO_2 and SO_2 from the post-combustion flue gas. The decay in sorbent reactivity for CO_2 capture over multiple reaction cycles poses a significant challenge to large-scale cyclic carbonation-calcination process. Thermal sintering at elevated temperatures and the subsequent loss in pore volume both contribute to decreasing CO_2 capture capacity. Intermediate hydration has been suggested to be one of the most promising approaches to overcome this limitation by regenerating the particle pore volume and surface area. However, most existing hydration studies have been performed at laboratory scale using thermogravimetric techniques, often investigating conditions not amenable to the 3-step CCR process. In addition, a reactor design specifically suited to high temperature steam hydration is warranted. Experimental data from such a reactor can offer insights to the mass transfer limitations associated with the gas-solid reaction of the extremely cohesive calcium hydroxide ($Ca(OH)_2$) particles under fluidized bed conditions. For the first time, this study reports the design of a bench-scale high-temperature steam hydrator for reactivation of CaO sorbent. The hydrator, consisting of a fluidized-bed reactor with additional internals such as an agitator and a cooling coil, was evaluated using cold-flow tests. Based on the results obtained, a high temperature reactor unit was constructed. The high-temperature steam hydration tests were performed at isothermal conditions in the range of 300-500 °C and 0.6-1.0 atm steam partial pressure. In particular, at an average temperature of 473 °C and P_{H2O} = 1 atm, the 30 min hydration conversion exceeded 70%. Sorbent reactivity toward CO_2 was substantially recovered by steam hydration and the correlation follows a linear trend. Lastly, the cooling coil design demonstrated a proof-of-concept for extracting the heat from this highly exothermic high temperature (>450 °C) reaction. This further expands the application of such reactor design beyond CO_2 capture, for other applications such as chemical heat pumps.

(774b) Dynamic Modeling of a Solid-Sorbent CO2 Capture System

Modekurti, S., West Virginia University
Bhattacharyya, D., West Virginia University
Zitney, S. E., National Energy Technology Laboratory
Miller, D. C., U.S. Department Of Energy, National Energy Technology Laboratory
Originally presented on: 11/8/2013 8:55:00 - 9:20:00

A one-dimensional (1D), two-phase, non-isothermal, dynamic model of a solid sorbent-based CO_2 capture system has been developed in Aspen Custom Modeler (ACM). The system consists of a two-stage bubbling fluidized bed (BFB) adsorber-reactor for CO_2 capture, a single stage moving bed regenerator for the sorbent regeneration, a multi-stage integral-gear centrifugal compressor system for CO_2compression, a pneumatic transport system for solids transport between the adsorber and regenerator, and the balance of the plant. In addition, a heat recovery system has been modeled for the regenerator which includes pre- and post-heat exchangers using steam/boiler feed water as the heating/cooling medium. The solid sorbent is NETL 32D developed at National Energy Technology Laboratory (NETL). The system of partial differential algebraic equations (PDAEs) has been generated using conservation equations, hydrodynamics, and reaction kinetics along with the appropriate boundary and initial conditions.

Flue gas from a supercritical pulverized coal (SCPC) plant enters at the bottom of the lower stage adsorber and regenerated sorbent enters the upper stage adsorber. Embedded heat exchangers have been considered for cooling of the bed using water as the cooling medium. The CO_2-rich sorbent is pneumatically transported to the regenerator through a pre-heater. An embedded heat exchanger in the regenerator maintains the unit at near-isothermal conditions. The regenerated sorbent passes through a low pressure steam generator that helps to recover heat for improving the overall efficiency. The medium selected for transporting the solid sorbent between the adsorber and regenerator is the one that minimizes the auxiliary power requirement without violating the overall target for CO_2 capture. The CO_2 rich gas stream from the regenerator is compressed to pipeline pressure by a multi-stage integral-gear compressor. A glycol absorption system is used to satisfy the pipeline specification for water in the CO_2stream. For off-design operation of the compressors, surge detection and control algorithms are developed.

The PDAE system with the appropriate boundary conditions is solved in ACM using the well-known method of lines which involves discretization of the spatial domain and dynamic integration of the time domain. The transient responses of various process variables such as CO_2 capture, flue gas and sorbent temperatures, and loading of the sorbent are studied by simulating typical disturbances such as change in the flue gas temperature, flowrate, and composition. A lower level control system is designed to maintain the overall CO_2 capture at the desired value in the face of such disturbances.

(774c) Comparative Environmental Life Cycle Assessment of CO2 Capture Systems Based On Ionic Liquid and Monoethanolamine Solvents

Farahipour, R.,
Karunanithi, A. T., University of Colorado Denver
Originally presented on: 11/8/2013 9:20:00 - 9:45:00

Carbon capture, storage and utilization (CCUS) is increasingly being viewed as a global climate mitigation option. Central to a viable CCUS system is the development of cost-effective methods for removal of CO_2 from flue-gas streams of power plants and other stationary CO_2 sources. In addition to economic viability it is important to consider the life cycle impacts of CCUS methods in order to evaluate the net benefits. Moreover, in addition to climate mitigation we would need to consider other issues such as increased life cycle impacts on human and/or ecosystem health. Amin-based (e.g., monoethanolamine) absorption systems for CO_2 capture is a fairly mature technology. However, they can be very energy intensive and less attractive from an environmental stand point due to emissions of the used volatile solvent components. In recent years, because of their low volatility, ionic liquids show significant promise as a CO_2 absorption solvent for post-combustion CO_2 capture. In this work, we present process simulation results of post-combustion CO_2 capture with monoethanolamine (MEA) and ionic liquid bmim [PF6]. The heat and material balances derived from these simulations are utilized to construct the life cycle inventory for the capture process. In addition data related to other upstream processes such as solvent (MEA and ionic liquid) production and downstream processes such as CO_2 compression, transportation, storage and/or utilization is used to perform a life cycle assessment with an aim to compare the environmental impacts of absorption systems based on these two different solvents.

(774d) Whole-Chain CCS System Modelling: Enabling Technology to Help Accelerate Commercialisation and Manage Technology Risk

Matzopoulos, M., Process Systems Enterprise Ltd.
Ramos, A., Process Systems Enterprise, Inc.
Originally presented on: 11/8/2013 9:45:00 - 10:10:00

The commercial implementation of CCS still faces significant challenges. Many of these arise from the fact that the whole chain – and, eventually, whole CO_2 transportation network – needs to be considered as a single system in order to make design and operation decisions that satisfactorily address the commercial imperatives and risk requirements of the various stakeholders along the chain. Even a quick analysis shows that design and operating decisions at the power plant can have a significant effect on storage providers at the other end of the chain, and vice versa. A systems modelling approach is essential, but there are currently no tools that can satisfactorily provide this capability over the whole CCS chain.

In order to address this, a £3m project was commissioned and co-funded by the ETI and project participants, who comprise E.ON, EDF, Rolls-Royce, Petrofac (via its subsidiary CO2DeepStore), Process Systems Enterprise (PSE) and E4tech. The project is aimed at delivering a robust, fully integrated tool-kit that can be used by CCS stakeholders across the whole CCS chain that will be released as a commercially-supported software product at the end of the project.

Because of the requirement for many stakeholders to be able to model areas of the chain beyond their own specific processes (for example, a key requirement for power generators is the ability to investigate the effects on their operation of amine and compression systems attached to their plants), the tool-kit will provide process models for the whole chain. This will also allow it to be used by groups who require a whole-chain view, such as engineering companies or government departments who need to quantify policy decisions.

The tool-kit includes models for conventional generation (pulverised coal and combined cycle gas turbine), new generation (gasification and oxyfuel), solvent-based carbon capture, compression, transmission and injection. The individual process models are mostly 'medium-to-high fidelity': multicomponent streams; equilibrium methods for vapour-liquid separation; comprehensive compressor models allowing multi-stage, multi-section compressors with manufacturers' curves; and distributed pipeline models for construction of pipeline networks that can take elevation into account. The capture models use rate-based techniques for accurate quantification of chemical and physical capture in order to quantify energy penalties accurately and allow meaningful analysis of transient operations.

Using the system modelling tool-kit it will be possible to look at single areas such as amine plants in detail; investigate partial-chain operations – for example, power generation, capture and compression; or analyse interactions across the whole chain and eventually the whole CO2 transmission network with multiple sources and multiple sinks.

The tool-kit also caters for interfacing of external software packages such as E.ON's PROATES, in order to simplify interoperability analysis while allowing companies to preserve existing workflows where necessary. In addition an advanced custom modelling capability means that it is easy to add models of new processes and combine them with existing flowsheeting components.

This presentation describes the status of developments to date in whole chain CCS system analysis and plans to roll the tool-kit software out to the wider 'CCS community' following the current testing phase.

(774e) Optimization of CO2 Capture, Utilization and Sequestration (CCUS) Supply Chain Networks

Hasan, M. M. F., Princeton University
Boukouvala, F., Princeton University
Floudas, C. A., Princeton University
Originally presented on: 11/8/2013 10:10:00 - 10:35:00

CO_2 capture, utilization and storage (CCUS) is an enabling technology toward reducing CO_2 emissions from stationary sources which include power plants, cement production plants, iron and steel plants, refineries, petrochemicals and gas processing plants. More than 60% of the total CO_2 emissions in the United States are attributed to stationary sources [1]. CCUS connects CO_2 sources and utilization/storage sites through three sequential operations: capture, compression, and transportation. CO_2 is first captured and compressed at the source plant, and then transported via pipeline to a site where it is utilized or injected for sequestration for geological storage.

It is critical to integrate capture, compression, transportation, utilization and sequestration in an optimal fashion. However, standalone development of CCUS components have made the design of large-scale CCUS networks a complex and challenging task. In this work, we present a mixed-integer optimization (MILP) model for the synthesis of CCUS supply chain networks for nationwide and regional CO_2 management. We design the CCUS supply chain network by considering simultaneous selection of source plants, capture technologies, capture materials, and locations of utilization and sequestration sites for the first time. We consider real geographic locations and reliable storage estimates from the available database. Furthermore, our cost model includes the costs of capture, compression, transportation and injection for sequestration, and also revenues from CO_2 utilization. The capture and compression costs are derived using advanced modeling, simulation and optimization techniques for a wide range of flue gas compositions and flow rates [2-3], which are crucial to incorporate stationary sources of different types in a large CCUS network. We apply our model to determine the CCUS supply chain network with minimum net expenditure for a given CO_2 reduction level.

Our model selects the optimal nationwide CCUS supply chain network from 3317 stationary sources, 491 saline formations, 1837 unmineable coal areas, and 161 utilization locations (oil and gas reservoirs). We also allow four leading capture technologies (absorption, membrane, pressure swing adsorption (PSA) and vacuum swing adsorption (VSA)), and 9 top materials (2 solvents, 4 zeolites [4-5], and 3 membranes) to choose from for CO_2 capture. Results suggest that it is possible to capture 50% of the current CO_2 emissions from the stationary sources in the United States at a cost less than \$30 for each ton of CO_2. Results for the CCUS in South Central United States, and Texas are also favorable. We demonstrate that it is crucial to consider both material and technology selections as integral parts of a large-scale CCUS network design.

References:

[1]. DOE/NETL. Carbon Sequestration Atlas of the United States and Canada, 2012.

[2]. Hasan, M. M. F.; Baliban, R. C.; Elia, J. A.; Floudas, C. A. modeling, simulation, and optimization of postcombustion CO_2 capture for variable feed concentration and flow rate. 1. Chemical absorption and membrane processes. Ind. Eng. Chem. Res. 2012a, 51, 15642-15664.

[3]. Hasan, M. M. F.; Baliban, R. C.; Elia, J. A.; Floudas, C. A. modeling, simulation, and optimization of postcombustion CO_2 capture for variable feed concentration and flow rate. 2. Pressure swing adsorption and vacuum swing adsorption processes. Ind. Eng. Chem. Res. 2012b, 51,15665-15682.

[4]. Hasan, M. M. F.; First, E. L.; Floudas, C. A. Cost-effective CO_2 capture based on in silico screening of zeolites and process optimization. Submitted for publication, 2013.

[5]. U.S. Provisional Patent Application #61/761,436 and #61/765,284.

(801a) Thermodynamic Analysis of Amine Solvents for CO2 Scrubbing

Mathias, P. M., Fluor Corporation
Reddy, S., Fluor Corporation
Smith, A. R., Fluor Corporation
Afshar, K., Fluor Corporation
Originally presented on: 11/8/2013 12:30:00 - 12:55:00

Evaluation of amine solvents for CO_2 scrubbing is difficult because superior performance is a complex combination of solvent characteristics (e.g., absorption capacity, heat of absorption, mass transfer and kinetics, hydraulic properties) and process innovations (e.g., absorber intercooling, lean-vapor compression, interheated strippers). In this work we tackle an important subset of the problem that is amenable to immediate objective analysis, i.e, what does solution thermodynamics teach us about relative solvent performance? We approach the problem by systematically creating solvents for which the only restriction is thermodynamic consistency, and then performing a relative evaluation using process simulation in a standard flowsheet. The study provides useful insight into the interplay between absorption capacity and enthalpy of solution. Our conclusion is that an optimum solvent is one that has mid-range enthalpy of solution and a CO_2 capacity that is just right (not to small and not too large) for the particular enthalpy of solution.

Of course, commercially successful solvents require other positive characteristics, like fast mass transfer and kinetics, low viscosity, high stability, and low toxicity. Comprehensive evaluation of new CO_2-capture solvents is necessary, but our study offers an important first step.

(801b) Thermodynamic Modeling of Carbon Dioxide Solubility in Aqueous Morpholine Solutions

S. Matin, N., Center for Applied Energy Research, University of Kentucky
Remias, J. E., Center for Applied Energy Research, University of Kentucky
Neathery, J. K., University of Kentucky
Liu, K., University of Kentucky

Originally presented on: 11/8/2013 12:55:00 - 13:20:00

Thermodynamic Modeling of Carbon Dioxide Solubility in Aqueous Morpholine Solutions

Naser S. Matin, Joseph E. Remias, James K. Neathery and Kunlei Liu*

Center for Applied Energy Research, University of Kentucky, Lexington, KY 40511, United States

Abstract

Employing the experimental CO_2 solubility in aqueous solutions of morpholine (6.14, and 8.85 m), a rigorous thermodynamic modeling is introduced. Both total pressure and a gas chromatography method were applied to determine CO_2 partial pressure over the solutions under study. In addition to solubility data the experimental speciation data was also included in system thermodynamic modeling. The Electrolyte-NRTL equations along with the needed thermodynamic equations were used to model the experimental CO_2 solubility and morpholine speciation data. Using this regression, the molecule-molecule and molecule-electrolyte binary parameters necessary for thermodynamic modeling of the system were determined. Using the CO_2 solubility data, the CO_2 heat of absorption in the solutions was also calculated. The regressed parameters from the experimental data yielded good agreement between the experimental and model values. A temperature dependent function for the morpholine carbamate stability constant was introduced, which displays a decreasing trend for carbamate formation with increasing temperature. The calculated morpholine carbamate stability constant is in acceptable agreement with literature. Considering the cyclic and non-cyclic secondary alkanolamines, morpholine shows a stronger carbamate bond as evidenced by its higher carbamate equilibrium constant compared to that of a linear secondary amine (DEA). This is partially reflected in the higher rate for CO_2 absorption in aqueous solutions of morpholine compared to the linear amine. The heat of absorption for morpholine was determined as a function of temperature and CO_2 solution loading. The temperature dependent Henry's law constant and heat of dissolution (physical solubility) for CO_2 in pure morpholine were also introduced. The determined CO_2equilibrium properties are in good agreement with most of the experimental data.

Keywords: Henry's constant; carbon dioxide; morpholine; vapor liquid equilibrium; modeling,

(801c) Modeling of CO_2 Absorption Into Aqueous Piperazine/Aminoethylpiperazine

Du, Y., The University of Texas at Austin
Rochelle, G. T., The University of Texas at Austin
Originally presented on: 11/8/2013 13:20:00 - 13:45:00

A novel blend of piperazine (PZ) with aminoethylpiperazine (AEP) has been proposed as a superior solvent for CO2 capture from coal-fired flue gas. Blending with AEP can remediate the precipitation issue of concentrated PZ while maintaining its high CO2 absorption rates, and high resistance to degradation. A rigorous model accurately predicting thermodynamic and kinetic properties for CO2 absorption in PZ-AEP-H2O-CO2 aqueous amine solutions has been developed in Aspen Plus® using the electrolyte-Nonrandom Two-Liquid (e-NRTL) activity coefficient model. Unavailable thermodynamic parameters of AEP-related species, including AEP, AEPH+, AEP(H+)2, AEPCOO-, AEP(COO-)2, H+AEPCOO-, (H+)2AEP COO- and H+AEP(COO-)2, were estimated by built-in models in Aspen Plus®, or by referring to corresponding PZ-related species as the starting point, and then sequential regressions were applied using related experimental data. The amine volatility data and heat capacity data for unloaded AEP solutions were regressed to determine Henry's constants and NRTL parameters for the AEP/H2O pair. The vapor-liquid equilibrium (VLE) data for AEP-H2O-CO2 were regressed to determine the standard-state properties (free energy of formation and heat of formation) and e-NRTL parameters of the AEP/AEP ion/H2O pairs. The VLE data for PZ-AEP-H2O-CO2 were used to identify the e-NRTL interaction parameters for the PZ-AEP pairs. The prediction of speciation by the model was validated by NMR measurement. The heat capacity and heat of absorption for PZ-AEP-H2O-CO2 were also predicted using this model. At 40 °C the heat of absorption of 6 m AEP is about 90–55 kJ/mol CO2 at operation loading range (0.27–0.33) and the heat of absorption of 5 m PZ/2 m AEP is about 78–70 kJ/mol CO2 at operation loading range (0.3–0.38). A steep decrease of heat of absorption was found at loading near 0.3 for 6 m AEP and at loading near 0.35 for 5 m PZ/2 m AEP.

(801d) Simulation of a Phase Change System for CO2 Capture

Pinto, D. D. D., NTNU
Monteiro, J.,
Knuutila, H., Norwegian University of Science and Technology
Tobiesen, F. A., SINTEF Materials and Chemistry
Mejdell, T., SINTEF Materials and Chemistry
Svendsen, H. F., Norwegian University of Science and Technology
Originally presented on: 11/8/2013 13:45:00 - 14:10:00

Chemical absorption is the most studied and applied process for carbon capture. However, it is yet an expensive process, mainly due to the regeneration section. Monoethanolamine (MEA) is still the benchmark solvent for CO_2 capture; nevertheless, new solvents and plant design could reduce the operational costs and make the process cheaper and industrially feasible. For instance, Rochelle et al. (2011) showed a new plant configuration that uses a two-stage flash to regenerate an aqueous piperazine (PZ) solution loaded with CO_2.

Following the trend of finding new solvents or solvent blends, an aqueous blend of Diethylaminoethanol (DEEA) and N-methyl-1,3-propanediamine (MAPA) at certain concentrations, herein called the blended system, was found to split into two liquid phases upon CO_2 loading. The upper phase is rich in DEEA and lean in CO_2 while the lower phase is rich in MAPA, CO_2 and water. Preliminary screening tests showed a high initial CO_2 absorption rate and a high capacity for a mixture of 5 M DEEA and 2 M MAPA.

Liebenthal et al. (2012) showed a simulation of a post combustion CO_2 capture process that uses the blended DEEA/MAPA system as a solvent in an integrated overall process. The results of the simulations showed a potential for reducing the reboiler temperature and duty compared to MEA. Moreover, the system was able to operate at higher desorber pressures than the conventional process which can benefit the compression section.

A simulation of a full capture process based on a blended system is presented in this work. The model was implemented in CO2SIM, an in-house SINTEF/NTNU simulator for CO_2 capture and a full pilot plant was simulated. The column modules are rate-based. This implementation was successfully applied in Knuutila et al. (2011) and Kvamsdal et al. (2011) where more information on the implementation of CO2SIM and CO2SIM dynamics can be found.

The model is based on VLE, kinetics and physical property data from the laboratory. The equilibrium behavior of the systems is based on a developed "soft" model which calculates total pressure and partial pressure of CO_2 as a function of temperature and loading. The model was validated by experimental data. A similar correlation was presented by Brúder et al. (2011). The "soft" model approach is a fast way to evaluate novel solvent systems and determine the best operational conditions for pilot plants (Kvamsdal et al., 2011).

To validate the simulation, a pilot campaign in the NTNU facilities was run with the blended system. The pilot was previously used for testing single phase amine and amino acids solutions (Aronu et al., 2010; Knuutila et al., 2011; Tobiesen et al., 2007). However, after minor changes, the plant was able to run the phase change system.

Since the pilot operates in a closed setup, after the injection of a desired amount of CO_2, the stripped CO_2 from the desorber was used as feed gas while the reboiler duty was varied. Additional injections of CO_2 were done so the amount of CO_2, and therefore, the loading was varied during the campaign.

Samples from the lean amine entering the absorber, rich amine samples after the separator and lean amine after the reboiler were withdrawn. After being centrifuged, to speed up the separation between the phases, the samples were analyzed for amine and CO_2 content. Additionally, the volume ratios (upper phase/lower phase) were measured.

The campaign was able to generate overall mass transfer coefficients values(Kg overall), CO_2 capture efficiencies, temperatures and pressure profiles for both the absorber and the desorber in addition to the heat duties required.

In this work, the model implemented in CO2SIM, after validation with the existing experimental data, was used to simulate a complete industrial plant for CO_2 capture. The model will be compared to experimental data for heat requirement, absorption rates and temperature profiles in the absorber and stripper over different operational conditions. Based on the validation energy consumption of a full scale plant will be simulated and the main advantages and limitations of blended, two phase systems are discussed. Moreover, the best estimated operational conditions will be presented and compared with a 30 weight % MEA campaign.

References

Aronu, U.E., Svendsen, H.F., Hoff, K.A., Knuutila, H., 2010. Pilot Plant Study of 3-(methylamino)Propylamine Sarcosine for Post-combustion CO_2 Capture, in: Farid, B., Fadwa, E. (Eds.), Proceedings of the 2nd Annual Gas Processing Symposium. Elsevier, Amsterdam, pp. 339-348.

Brúder, P., Grimstvedt, A., Mejdell, T., Svendsen, H.F., 2011. CO_2 capture into aqueous solutions of piperazine activated 2-amino-2-methyl-1-propanol. Chemical Engineering Science 66, 6193-6198.

Knuutila, H., Aronu, U.E., Kvamsdal, H.M., Chikukwa, A., 2011. Post combustion CO_2 capture with an amino acid salt. Energy Procedia 4, 1550-1557.

Kvamsdal, H.M., Haugen, G., Svendsen, H.F., Tobiesen, A., Mangalapally, H., Hartono, A., Mejdell, T., 2011. Modelling and simulation of the Esbjerg pilot plant using the Cesar 1 solvent. Energy Procedia 4, 1644-1651.

Liebenthal, U., Pinto, D.D.D., Monteiro, J.G.M.-S., Svendsen, H.F., Kather, A., 2012. Overall Process Analysis and Optimisation for CO_2 Capture from Coal Fired Power Plants based on Phase Change Solvents

Forming Two Liquid Phases, GHGT11. Energy Procedia, Kyoto, Japan.

Rochelle, G., Chen, E., Freeman, S., Van Wagener, D., Xu, Q., Voice, A., 2011. Aqueous piperazine as the new standard for CO2 capture technology. Chemical Engineering Journal 171, 725-733.

Tobiesen, F.A., Svendsen, H.F., Juliussen, O., 2007. Experimental validation of a rigorous absorber model for CO_2 postcombustion capture. AIChE Journal 53, 846-865.

(801e) Solubility of Carbon Dioxide in Aqueous Solutions of Potassium Carbonate ("Activated Hot Potassium Carbonate Process")

Wagner, M., University of Kaiserslautern
McCann, N., University of Kaiserslautern
Kumelan, J.,
Speyer, D.,
Maurer, G., Laboratory of Engineering Thermodynamics
Hasse, H., University of Kaiserslautern
Originally presented on: 11/8/2013 14:10:00 - 14:35:00

In the Hot Potassium Carbonate Process carbon dioxide is removed from gaseous streams by chemical absorption in an aqueous solution of potassium carbonate. It is common to "activate" the solvent by adding so-called "promoters", like borates and vanadates, the role of which is, however, not fully understood. In the present investigation we studied the influence of borates and vanadates on the equilibrium solubility of carbon dioxide in aqueous solutions of potassium carbonate in three steps. In the first step the complex reaction network in these solutions was studied by ^{51}V-NMR-spectroscopy. The investigations reveal that besides the various complexes that are described in the literature, carbonato-vanadate complexes exist. Based on these experimental results and on literature data, the reaction network was modeled quantitatively [1]. In the second step, the solubility of carbon dioxide in four different activated aqueous solutions of K_2CO_3 was determined experimentally at two temperatures that are typical for carbon dioxide absorption (343 K) and solvent regeneration (383 K) in the Hot Potassium Carbonate Process. The mass fraction of K_2CO_3 in the solvent was 0.26 g/g, the mass fractions of boron (vanadium) was varied between 0.006 and 0.013 g/g (0.01 and 0.02 g/g). Two set-ups were used for the experiments: A headspace gas chromatographic technique was applied to determine the solubility of CO_2 at partial pressures of carbon dioxide between (1 and 140) kPa and the synthetic gas solubility technique was applied for total pressures between (0.4 and 10) MPa. The new experimental results are compared to predictions from a physico-chemical model for the solubility of CO_2 in aqueous solutions of K_2CO_3 [2]. The results reveal that both the borate and the vanadate reduce the solubility of CO_2 in such solutions. In the third step, the model was extended to describe that "salting-out" phenomenon quantitatively, taking into account not only the new gas solubility data but also the new data on the chemical reactions [1]. The new model provides an excellent basis for simulations of the Hot Potassium Carbonate Process.

[1] McCann N., Wagner M., Hasse H.; Dalton Trans., 2013, **42**, 2622-2628

[2] PÃ©rez-Salado Kamps, Á., Meyer E., Rumpf B., Maurer, G.; J. Chem. Eng. Data, 2007, **52**, 817-832

(808a) CO_2 Capture By Gas Hydrate Formation: Combined Effect of Thermodynamic and Kinetic Promoters

Daraboina, N., Denmark Technical University
von Solms, N., Technical University of Denmark (DTU)
Originally presented on: 11/8/2013 12:30:00 - 12:55:00

Carbon capture by gas hydrate crystallization is a promising technology currently being considered in the battle to mitigate climate change. However, the requirement of both high pressure and slow formation rates are serious limitations which need to be resolved prior to implementing this technology at demonstration scale. Various additives (e.g.: THF, TBAB, cyclopentane and others) are known to promote hydrate formation thermodynamically by reducing the hydrate formation pressure. Furthermore there are also kinetic promoters such as surfactants which can increase the hydrate crystallization rate based on the concentration. The appropriate combination of thermodynamic and kinetic promoters has the potential to move the technology towards practical application.. In this work, the hydrate-based process for carbon dioxide capture was tested in a rocking cell (RC-5; Systemtechnik, Germany) apparatus. Various combinations of surfactant (kinetic promoters: SDS and CTAB) and THF (thermodynamic promoter) in liquid water were assessed in order to find the optimum operating conditions.

(808b) CO2/Brine/Rock Interactions Under CO2 Sequestration Conditions

Soong, Y., National Energy Technology Laboratory (NETL), Office of Research and Development, Department of Energy

Howard, B. H., National Energy Technology Laboratory

Hedges, S. W., U.S. Department of Energy

McIntyre, D., National Energy Technology Laboratory

Warzinski, R. P., U.S. Department of Energy, National Energy Technology Laboratory

Irdi, G.,

Haljasmaa, I.,

Crandall, D., National Energy Technology Laboratory (NETL), Office of Research and Development, Department of Energy

Originally presented on: 11/8/2013 12:55:00 - 13:20:00

CO_2/brine/rock interactions under CO_2 sequestration conditions

Y. Soong*, B. H. Howard, S. W. Hedges, D. McIntyre, R. Warzinski, G. Irdi, I. Haljasmaa[1], and D. Crandall[1]

U.S. DOE, National Energy Technology Laboratory,

P.O. Box 10940, Pittsburgh, PA 15236

[1] URS Co.

It has indicated by Intergovernmental Panel on Climate Change Study that injection of CO_2 into confined geological formations, given their potentially massive carbon storage capacity and widespread geographic distribution, represents one of the most promising options for mitigating anthropogenically-induced global climate change. The Mount Simon sandstone in the Midwest region of the U.S is recognized as a promising candidate host reservoir for carbon sequestration. The Mount Simon is a deep saline aquifer and is a primary reservoir target for large scale carbon dioxide injection tests due to its proximity to CO_2 sources, favorable depth, thickness, permeability, porosity, and the presence of the overlying Eau Claire Formation as a seal. The Midwest Geological Sequestration Consortium (MGSC) has selected the Mount Simon as the reservoir at its major demonstration project. While the Mount Simon Sandstone has ideal reservoir characteristics in some areas, there are significant variations in porosity, permeability, and mineralogy.

Numerous studies have been conducted to investigate changes in host rock properties when exposed to CO_2. Luquot and Gouze conducted CO_2-enriched fluid flow through a carbonate core sample under in situ sequestration conditions (T = 100 °C and P = 12 MPa). Under these conditions, a decrease in permeability and porosity is seen that is linked to precipitation of Mg-rich calcite. Gouze and Luquot also conducted an X-ray microtomography characterization study of porosity, permeability and reactive surface changes during the dissolution of pure calcite in a brine-CO_2 mixture. It showed that the increase in permeability is due to the decrease of the tortuosity for homogeneous dissolution, whereas it is due to the combination of tortuosity decrease and hydraulic radius increase for heterogeneous dissolution. Jove Colon et al., studied the permeability, porosity and reactive surface area evolution during dissolution of nonfractured, clay-free Fontainebleau sandstone cores using a flow-through percolation reactor. The experiments were performed at 80°C with 0.1 M NaOH solution. The permeability was found to vary depending on where the dissolution occurred: in the pore void or the pore throat. Liu et al., conducted a simulation study of CO_2 sequestration in the Mount Simon sandstone via TOUGHREACT. They indicated that a strongly acidified zone (pH 3-5) forms in the areas affected by the injected CO_2 and consequently causes extensive secondary mineral precipitation (calcite, magnesite, ankerite, alunite and anhydrite) and dissolution of feldspars. Thus, the injection of CO_2 into a deep saline aquifer affects porosity since it can cause both mineral dissolution and precipitation in the formation. To predict if there will be any changes to the reservoir during sequestration, it is critical to understand the behavior of the authigenic minerals that line pore spaces when they are in contact with the CO_2 saturated brine.

An experimental study was conducted to assess the impacts of injected CO_2. Core samples from the Mount Simon formation were placed in CO_2 saturated brines at sequestration temperature of 85 °C and pressure of 23.8 MPa in the presence of synthetic Illinois Basin brine for a period of six months in a set of 1.3 liter high-pressure vessels (17-4PH-1150 stainless steel, 10.2 cm I.D. x 16.5 cm depth, manufactured by Thar Technologies, Inc.). One sandstone sample with a porosity of 7.9% was obtained at depth of 1,769.7 m from a well located in Vermillion County, IN. The other sample with a porosity of 1.4% was collected at depth of 2634.2 m from a well located in Knox County, IN. The major mineral composition for the first sample is 78% quartz, 15% Feldspar and with less than 2% microcline and trace minerals. The primary mineral composition of the second sample obtained from Knox County is 70% quartz, 22% Feldspar, and around 4% illite/muscovite. Micro CT, CT, XRD, SEM, petrography, porosity, and permeability analyses were conducted prior to the six month exposure. Upon completion of the experiment, the analyses were repeated.

The permeability of the Vermillion sandstone after the six month exposure to CO_2 and brine was 0.85 mD compared to 1.6 mD for the fresh sample. On the other hand, the permeability of Knox sandstone was 69 nD compared to 50 nD for the fresh sample. The results suggest that the exposure altered the pore characteristics and thus the permeability of the sample. Micro CT analysis as well as SEM backscattered images were used to probe the Vermillion and Knox Co. sandstone before and after the six month exposure to the CO_2/brine environment. It can be seen that mineral dissolution and mineral precipitation occurred after the samples were exposed to CO_2/brine for six months. Mineral precipitation occurred in the Vermillion sample pores and cracks blocking flow and resulting within the observed 50% decrease in permeability. For the case of Knox sandstone, the mineral dissolution occurred in pore though the precipitation was observed in the exterior of the sample. This might result in increasing in permeability. In addition, from the analyses of the reacted brines within the reactor, it indicated the possibly of Feldspar dissolution.

When CO_2 is injected into a saline formation, acidic conditions form in the brine in the nearby portion of the reservoir thus reducing the pH of the brine in the vicinity of the injection system. Minor minerals in the sandstone, such as K-feldspar, calcite, kaolinite, microcline etc., that are stable in the native environment but not stable at lower pH can dissolve. Eventually, the increase of cation concentration produced by mineral dissolution can supersaturate the fluids with respect to other minerals. If this occurs, precipitation of new minerals can result in a change of porosity and permeability which could affect both CO_2 storage capacity and fluid movement. The issue is further complicated by the evolution of pore geometry due to these geochemical reactions that may affect the overall hydrologic properties of the host rock and therefore the performance of the unit as an effective reservoir.

The decreasing permeability found in Vermillion sandstone was probably due to Feldspar dissolution, migration and precipitation altering the sandstone pore/crack structure. For the case of sandstone from Knox county, the permeability increased by 38%. It might be due to the Feldspar dissolution in the pore and mineral precipitation at exterior surface of the sandstone. Investigations are continuing to more clearly elucidate the mechanisms and geochemistry of this system, especially at the pore structure scale.

(808c) Novel CO2 Capture and Carbonation Process With Waste Concrete

Shuto, D.,
Yamasaki, A., Seikei University
Hongo, T., Seikei University
Iizuka, A., Tohoku University
Originally presented on: 11/8/2013 13:20:00 - 13:45:00

A novel CO_2 capture and carbonation process is developed. In this process, wastes containing calcium, such as waste cement particles or steel slug, are used as a calcium source for carbonation reaction with CO_2 emitted from point sources. The process is composed of several steps. CO_2 in the flue gas is captured through absorption process with alkaline solution such as sodium hydroxide to form sodium carbonate and sodium bicarbonate. The calcium content in the waste cement particles is extracted with acid such as nitric acid or acetic acid to form solution containing calcium. Mixing of these two solutions leads to the precipitation of calcium carbonate: CO_2 fixation via carbonation reaction. Electrodialysis is applied for the regeneration of acid and alkaline from the remained solution after filtration of the precipitant. Thus, the net input of the process is CO_2 in the flue gas and waste cement particles, and the output is calcium carbonate. The process feasibility was examined through laboratory -scale experimental studies under various operation conditions. Based on the experimental results, the process was optimized and evaluated in terms of the power consumption and cost for the CO_2 recovery, and compared with conventional processes. Net CO_2 sequestration can be achieved by the process under certain conditions.

(808d) Direct Capture and Conversion of CO_2 From Flue Gas Via Ex-Situ Mineral Carbonation Using Chemical and Biological Catalysts

Swanson, E. J., Columbia University
Brady, P. V., Sandia National Laboratories
Park, A. H. A., Columbia University
Originally presented on: 11/8/2013 13:45:00 - 14:10:00

Converting carbon dioxide from flue gas directly into a solid magnesium carbonate presents a compelling option for the mitigation of power plant emissions. The carbonate product is environmentally benign and thermodynamically stable, thus creating a safe and reliable alternative to geological storage. However, given the low partial pressure of carbon dioxide in flue gas and the low reactivity of magnesium bearing minerals, achieving the kinetics required to make such a process feasible is challenging. In addition, magnesium carbonates formed at low temperature tend to incorporate significant waters of hydration, which increases transportation requirements and limits utilization of by-products. Each of these issues was addressed in the design of a multistep process for the conversion of magnesium silicates to carbonates as a carbon capture and storage option. The mineral dissolution reaction is accelerated using a mixture of organic chelating agents to produce a concentrated magnesium solution. This acceleration appears to occur through a stabilization effect on the pores formed as the silicate dissolves. The concentrated magnesium solution produced during the dissolution of the silicate is used to drive the precipitation of carbonates directly from flue gas. Through strict control over the chemistry of a simultaneous absorption and precipitation process, the phase of magnesium carbonate formed can be controlled. In addition, the mass transfer rate of carbon dioxide from flue gas into the precipitating liquor is accelerated using biological catalysis, which is shown to provide significant savings in process equipment size and operating parameters. Finally, the process as a whole is demonstrated to have a negative overall carbon balance through life cycle assessment.

(808e) A Green Approach to Mitigate CO2 Using Mixed Microbial Culture

Raghuvanshi, S., Birla Institute of Technology & Science, Pilani
Gupta, S., Birla Institute of Technology & Science, Pilani
Mishra, S.,
Originally presented on: 11/8/2013 14:10:00 - 14:35:00

Carbon dioxide is one of the GHG's (Green House Gases) & responsible for Global warming.. As of July 2011, CO_2 in the earth's atmosphere is at a concentration of 392 ppm by volume which is 35% higher as compared to the mid-1800s. Carbon dioxide, which is estimated to contribute 50-60% of the Global warming in the last century, is majorly released during the combustion of fossil fuel. The CO_2 concentration in earth atmosphere can be reduced by biological, chemical and technological techniques available through either mitigating or sequestering emissions. In the recent years, biological route has been identified as one of the most viable options available for the mitigation of CO_2. There are six carbon fixing pathway's. Out of which, the Calvin cycle (Calvin-Benson-Bassham) is used by photoautotroph's to fix CO_2. There are microbes which are ubiquitous in nature and thrive in extreme environment classified as chemoautotrops. They also classified as Extremophiles. The requirement of industries is to develop a economic, efficient and effective system which not only fix CO_2 but also capable to withstand in the severe industrial environment. Mixed culture is proved to be a superior candidate, as the consortium of different engineered species develop a symbiotic relationship which helps each other not only to survive in harsh conditions but also curb down CO_2. The another possibilities are simultaneous waste water treatment, production of Methane, and bio-hydrogen production. The present work deals in identification of microbes present in sludge of waste water treatment plant having CO_2 fixation capabilities. It also gives the insight of the relation which exists between natural selected mixed culture and to understand the capabilities of CO_2 fixation. The batch study has been conducted to measure the capabilities of mixed culture to mitigate CO_2 for five different V/V % concentration of CO_2 4, 12, 16, 22 and 25. It was observed that CO_2 is completed consumed by microorganisms in 4 days. Mitigation rate for different concentration of CO_2, bio-mass growth rate and resulting product has been analyzed on per hour basis and per day basis. Mixed culture is found to be superior candidate than photoautotrophic organisms for CO_2 mitigation. They show a promising future to develop a low cost, compact size closed system for *in-situ* CO_2 mitigation.

(816a) Steady-State and Dynamic Modeling of a Moving Bed Adsorber/Regenerator for Solid Sorbent CO2 Capture

Kim, H., National Energy Technology Laboratory
Modekurti, S., West Virginia University
Miller, D. C., U.S. Department Of Energy, National Energy Technology Laboratory
Bhattacharyya, D., West Virginia University
Zitney, S. E., National Energy Technology Laboratory
Originally presented on: 11/8/2013 12:30:00 - 12:50:00

Moving bed reactors hold strong promise for solid sorbent-based CO_2 capture processes by enabling larger working capacities than other reactor types such as fixed- and fluidized-bed reactors. A first-principles model of a moving bed reactor can be used to investigate tradeoffs between various design alternatives and determine the optimal configurations, process conditions, and sorbent types. With this motivation, one-dimensional (1D), two-phase, non-isothermal, steady-state and dynamic models of a moving bed reactor have been developed for use in the U.S. Department of Energy's Carbon Capture Simulation Initiative (CCSI). The models are flexible over a wide range of operating conditions such as gas and solids flowrates and compositions, and operating temperatures. They can be used to represent both an adsorber for CO_2 capture and a regenerator for removing CO_2 from solid sorbents.

In these moving bed reactors, solid sorbent particles move downward due to gravity, contacting a rising gas stream in a counter-current configuration. Since the adsorption reaction is exothermic, an immersed cooler is considered in the adsorber for enhancing solid sorbent loading. An immersed heater is considered in the regenerator to maintain a near-isothermal operating condition. An integral heat-recovery system is used for recovering heat from the hot sorbent leaving the regenerator. The recovered heat is then used to preheat the solid sorbent from the adsorber. The reaction kinetics, heat and mass transfer, and the hydrodynamics are considered in the beds and in the heat recovery system for the regenerator. Sorbent NETL32D, a solid sorbent developed by NETL, is modeled in this work. The kinetic model considers three reactions and has been developed by lumping the internal mass transfer resistances. To capture the mass transfer resistance from/to the bulk to/from the solid sorbent surface, especially in the regenerator that operates at higher temperature, external mass transfer resistances have been modeled. The hydrodynamics of moving beds have not been widely studied, so most coefficients have been derived by analogy with correlations for fixed and fluidized bed systems. The gas flow is modeled by convective gas movement with axial dispersion. For modeling heat exchange between the moving bed and the embedded heat exchangers, heat transfer coefficients have been derived by considering an enhanced packet renewal theory.

The system of partial differential algebraic equations with the appropriate boundary conditions is solved in Aspen Custom Modeler (ACM) using the well-known method of lines. The effect of operating conditions and design parameters on the performance of the adsorber/regenerator reactor has been studied using the steady-state model. Using the dynamic model, a number of disturbances are simulated to study the transient response of the moving bed adsorber/regenerator.

(816b) Modeling, Analysis and Optimization of Process Technologies for Flue Gas Dehydration

Gao, H., Northwestern University
You, F., Northwestern University
Snurr, R., Northwestern University
Originally presented on: 11/8/2013 12:50:00 - 13:10:00

Carbon Capture and Sequestration (CCS) of flue gas from large power plants have received increasing attention because of concerns about the role of CO_2 in global climate change. Among different potential technologies for CCS, adsorption processes based on metal-organic frameworks (MOFs) have received significant attention because MOFs can be designed to achieve very high selectivity of CO_2 over N_2[1]. However, the stability and capacity of MOFs can be seriously affected by the presence of water in the flue gas[2]. Many researchers are working to develop new MOFs that can operate in the presence of water, but here we explore options for removing water from flue gas before a downstream adsorption process. Such a dehydration step can contribute a significant part of the total cost for CO_2 capture[3]. Furthermore, different MOFs may have different tolerance of moisture, so the dehydration requirement (i.e. the water vapor percentage in the dried flue gas), needs to be considered as a variable in order to be connected with possible material selection for the downstream process. To reduce the cost of an adsorption based carbon capture process, the dehydration unit needs careful design and optimization, but this issue is rarely addressed in the existing literature. In this study, we modeled and optimized different process technologies for the dehydration of flue gas. Our goal is to identify the best dehydration technology and minimize the cost of the dehydration process.

We investigated four alternatives for flue gas dehydration, including compression & cooling, cooling & condensation, TEG absorption and solid adsorption. They are compared in terms of cost and separation performance (i.e. loss of other components when satisfying the dehydration requirement). We varied the dehydration requirement from 0.1% to 1% of water in the dry gas for all four technologies. The first three processes are continuous processes and can be well simulated using Aspen Plus. We then apply a response surface method (RSM)[4, 5] to build a function linking the input-output data and minimize the process costs. Adsorption is a dynamic process operated in a cyclic manner[6]. Governing equations form a partial differential equation system, which is fully discretized both in temporal and spatial domain. We apply a first order finite volume method to discretize the spatial domain, and a Radau collocation scheme for temporal discretization. This transfers the original dynamic optimization problem to a nonlinear optimization problem, which can then be modeled and solved in GAMS using general-purpose MINLP solvers to obtain the optimal operating conditions and cost.

Results show the cost for water removal from flue gas can range from a few dollars to tens of dollars per ton CO_2 captured, varying with the dehydration requirement. Among the four technologies, cooling &

condensation proved to be the cheapest process for water removal, but it can only dehydrate to about 0.5% water content with no more than 10% loss of CO_2. TEG absorption loses the least CO_2 but the process cost is higher. The advantages and disadvantages of other technologies are also analyzed and discussed. The consideration of different dehydration requirements provides the foundation for linking with material selection in the downstream adsorption process.

Reference

[1] M. Zhao, A. I. Minett, and A. T. Harris, "A review of techno-economic models for the retrofitting of conventional pulverised-coal power plants for post-combustion capture (PCC) of CO2," *Energy & Environmental Science,* vol. 6, pp. 25-40, Jan 2013.

[2] A. C. Kizzie, A. G. Wong-Foy, and A. J. Matzger, "Effect of Humidity on the Performance of Microporous Coordination Polymers as Adsorbents for CO2 Capture," *Langmuir,* vol. 27, pp. 6368-6373, May 17 2011.

[3] M. M. F. Hasan, R. C. Baliban, J. A. Elia, and C. A. Floudas, "Modeling, Simulation, and Optimization of Postcombustion CO2 Capture for Variable Feed Concentration and Flow Rate. 1. Chemical Absorption and Membrane Processes," *Industrial & Engineering Chemistry Research,* vol. 51, pp. 15642-15664, Dec 5 2012.

[4] J. A. Caballero and I. E. Grossmann, "An algorithm for the use of surrogate models in modular flowsheet optimization," *Aiche Journal,* vol. 54, pp. 2633-2650, Oct 2008.

[5] D. R. Jones, M. Schonlau, and W. J. Welch, "Efficient global optimization of expensive black-box functions," *Journal of Global Optimization,* vol. 13, pp. 455-492, Dec 1998.

[6] R. T. Yang, *Gas separation by adsorption processes*. Boston: Butterworths, 1987.

(816c) On the Design and Modeling of a Rotary Reactor for Chemical-Looping Combustion With a Nickel-Based Oxygen Carrier

Zhao, Z., Massachusetts Institute of Technology
Chen, T., Massachusetts Institute of Technology
Ghoniem, A. F., Massachusetts Institute of Technology
Originally presented on: 11/8/2013 13:10:00 - 13:30:00

Chemical-looping combustion (CLC) is a novel and promising option for carbon capture and sequestration (CCS). Currently almost all the research has been focused on developing CLC-based interconnected fluidized-bed reactors. In this report, a rotary reactor with a nickel-based oxygen carrier (OC) for gas fueled CLC is designed and analyzed. In the reactor, a solid wheel rotates between the fuel and air streams at the reactor inlet and exit. The rotary wheel consists of a large number of micro-channels with nickel oxide coated on the inner surface of the channel walls. The OC oxidizes the fuel in the fuel zone to generate undiluted CO_2 and then it is regenerated in the air zone. Thus pure CO_2 can be readily obtained from the flue streams of the fuel zone and the dilution from the air is avoided. Gas flows through the reactor at elevated pressure and it is heated to high temperature by fuel combustion which can then be used to drive the gas turbine. Design parameters have been selected on the basis of the characteristics of the nickel-based OC and for a thermal capacity of 1MW, the reactor has a diameter of around 1.0m, and a height of about 0.5m. Owing to a relatively slow oxidation rate, an intermediate range of the OC conversion as well as a relatively large share of air sector is used for the design. A one-dimensional plug-flow model has been developed to predict the performances of the reactor. The results indicate that both the fuel conversion and the carbon separation efficiencies are close to unity. In the periodic stationary operation, stable performances are obtained with the gas temperature increasing monotonically from 550℃ to 1200℃. The overall energy in the solid phase is balanced between the reaction heat release, conduction and convective cooling. Sensitivity study has been conducted on the base-case design and the results revealed that the most important parameters are the reduction kinetics, the operating pressure and the feed stream temperatures. Advantages of the rotary design with the nickel-based OC include the compactness, up-scaling feasibility, periodic and stationary operation with limited energy penalty from operation. The potential challenges are also discussed in the report.

(816d) Experimental Characterization of Rotary Reactor for Chemical-Looping Combustion

Chen, T., Massachusetts Institute of Technology
Uddi, M.,
Zhao, Z., Massachusetts Institute of Technology
Ghoniem, A. F., Massachusetts Institute of Technology
Originally presented on: 11/8/2013 13:30:00 - 13:50:00

Chemical looping combustion (CLC) is currently one of the most promising methods to achieve Carbon capture and sequestration (CCS). A novel rotary reactor concept for gas fueled combustion is proposed to eliminate the large energy penalties of traditional CCS technologies during the separation process.[i] The core of the reactor is a solid wheel consisting of numerous micro-channels rotating between fuel and air streams. Metal oxide, known as the oxygen carrier (OC), that reacts with fuel to generate undiluted CO_2 is coated on the inner surface of the channel walls. After the reduction reaction in the fuel zone, the OC is carried over to the air zone by the rotary wheel, where it is regenerated by the oxidation reaction. As a result pure CO_2 can be recycled and sequestrated from the flue streams of the fuel zone by condensing water.

A detailed numerical model has been developed to simulate and validate the capability of the reactor design to completely convert the fuel and capture CO_2 while maintaining desirable stabilities.[i],[ii] However, as the new reactor design employs different reaction mechanism than what is available in literature, experimental investigation is needed. A button cell reactor, which is connected to a mass spectrometer (MS) with real time gas analysis capability, is developed. The main goal of performing the experimental study is to investigate the physical and chemical processes that take place when reactive flow goes through the reactor. This includes fuel conversion, oxidation and reduction reactions of the OC, OC material selection and optimization, and the temperature variation and distribution along the reactor channel. Results of the experiments will be analyzed and used to validate the numerical simulations.

[i] Z.Zhao, T.Chen, A. F. Ghoniem. Rotary bed reactor for CLC with carbon capture. Part I: Reactor design and model development. Energy Fuels, 2013, 27 (1), pp 327–343

[ii] Z.Zhao, T.Chen, A. F. Ghoniem. Rotary bed reactor for CLC with carbon capture. Part II: Base case and sensitivity analysis. Energy Fuels, 2013, 27 (1), pp 344–359

(816e) Comparison of Various Carbon Capture Technologies to Reduce CO2 Emissions From a Cement Plant

Ozcan, D. C., University of Edinburgh
Bocciardo, D., University of Edinburgh
Ferrari, M. C., University of Edinburgh
Ahn, H., University of Edinburgh
Kierzkowska, A. M., ETH Zurich
MÃ¼ller, C. R., ETH Zurich
Brandani, S., University of Edinburgh
Originally presented on: 11/8/2013 13:50:00 - 14:10:00

Substantial reductions of global CO_2 emissions by at least 50% are required by 2050 according to Intergovernmental Panel on Climate Change [1]. While CO_2 reduction technologies have been mostly developed for the purpose of carbon capture from the power generation industry, it is important to extend this to the cement industry in order to achieve the CO_2 capture target projected over the next 40 years. The cement industry accounts for over 2 billion tonnes of CO_2 emitted per year [2,3]. Only marginal reductions in CO_2 emissions can be achieved through improving the energy efficiency of the cement manufacture process. Significant reductions can be obtained only if a carbon capture unit is integrated in a cement plant given that around 60% of CO_2 emission comes from the calcination of the raw material.

Both post-combustion and oxy-firing would be deemed appropriate for carbon capture from cement works. To compare the different configurations full process flowsheets have been developed and simulated in Unisim Design Suite from Honeywell. The process considered is based on the reference case from the IEA [4] and includes the main raw material decomposition and clinkerization reactions. The base case simulation was then modified to integrate three different post-combustion capture processes: amine-scrubbing; calcium looping process; and membrane-based capture units. Furthermore, an advanced configuration of Ca-looping process has been investigated where the energy intensive air separation unit was replaced with a chemical looping combustion (CLC) cycle. Each case has been optimised to minimise its energy consumption and compared in terms of CAPEX/OPEX and its resulting CO_2 avoided cost at the same capture rate.

[1] Intergovernmental Panel on Climate Change (IPCC) (2007b). Metz, Bert. Davidson, Ogunlade. Bosch, Peter. Dave, Rutu. Meyer, Leo (Eds.). *Climate Change 2007: Mitigation. Contribution of Working Group III to the Fourth Assessment Report of the Intergovernmental Panel on Climate Change.* Cambridge University Press, Cambridge, United Kingdom and New York, NY, USA.

[2] CW Group *Global Cement Volume Forecast Report*, 2012.

[3] IEA Carbon Emission Reductions up to 2050. In *Cement Technology Roadmap 2009*; International Energy Agency: Paris, 2009.

[4] IEA *CO_2 Capture in the Cement Industry*, July 2008/3, 2008.

(816f) Non-Isothermal Performance Modeling for CO2 Adsorption

Breault, R. W., National Energy Technology Laboratory
Huckaby, E. D., National Energy technology Laboratory
Shadle, L. J., National Energy Technology Laboratory
Spenik, J., REM, Corporation
Originally presented on: 11/8/2013 14:10:00 - 14:30:00

The National Energy Technology Laboratory as well as other institutions are in the process of conceptualizing new technology as an alternative to amine based absorption processes for CO2 capture from large sources such as utility power generation facilities. Many, but not all of these advanced dry processes are based upon sorbents composed of supported polyamines. This particular investigation gives a comparison between the predicted performance with a non-isothermal model and the experimental performance of one of these new supported polyamine sorbents. In the continuous process CO2 from is absorbed from flue gas in the absorber, a cyclone separates the carbonated particles from the de-carbonated flue gas and then regenerates the sorbent, creating a concentrated stream of pure CO2 for sequestration. In the current work, the reactors are run in a batch mode, that is without circulating solids. In this work, experimental data from adsorption and regeneration in a fluidized bed operated in batch mode are compared to simulation predictions from a 3-dimentional multiphase flow model complete with chemistry and thermal effects. The effects of gas flow rate and reactor temperature are explored and compared with experimental measurements. The specifics for the experimental facility and the model as well as the comparative analysis between the data and the simulation results will be discussed.

(816g) A Correlation for the Impact of Extracting Steam From Power Plants for Carbon Capture Processes

Berger, A. H., Electric Power Research Institute
Bhown, A. S., Electric Power Research Institute
Originally presented on: 11/8/2013 14:30:00 - 14:50:00

A Correlation for the Impact of Extracting Steam from Power Plants for Carbon Capture Processes

Adam H Berger and Abhoyjit S Bhown

Electric Power Research Institute (EPRI), Palo Alto, CA

Abstract

Post combustion carbon dioxide capture and storage (CCS) from power plants is an energy intensive process often consuming one quarter to one third of the total power generated by the base power plant. This parasitic energy is comprised of two main terms: the auxiliary load, which accounts for pumping, compression, and other mechanical or electrical work; and the thermal load, which is typically steam extraction for process heating. It is important to understand how these two terms interact with the base power plant so that accurate parasitic energy can be calculated and CCS materials or processes can be developed and compared based on actual impact on the power output of power plants.

While the impact of the auxiliary load on the power plant output is immediately apparent, the impact of the thermal load on the power plant is more difficult to quantify. The thermal load is used to provide heating energy to a CCS process in the form of steam extracted from the steam cycle of the power plant. The extracted steam is used to provide process heating such as in the reboiler of a solvent stripper or heating solid material in a temperature swing adsorption process. The electrical impact of extracting steam is defined as the power loss in the turbines from the decreased steam flow rates. This power loss is a function of the temperature and pressure of steam extraction as well as the quantity of steam extracted. The heating energy has components, the energy from de-superheating the steam extracted and energy from the heat of vaporization of steam condensing at the pressure of extraction. This condensation at the saturation temperature also determines the temperature at which process heating can be provided. In addition to the temperature and pressure of the steam extracted, the quantity of steam extracted is also important because the effect of steam extraction on low-pressure turbine operation and efficiency.

In this work, we use state-of-the-art models of low pressure turbines in power plant steam cycles to quantify the effect of steam extraction on power plant operation. We further develop a correlation for the impact on a power plant of providing thermal energy at a specified quantity and temperature to a CCS process. This correlation can be used in any process or material development to more accurately compare the effect of different process configurations or material properties on the total energy impact of CCS and thus compare trade-offs between mechanical work and thermal load or between capital and operating expenses.

AUTHOR INDEX

AUTHOR INDEX

AUTHOR INDEX